Oh Yes You Can
GET BETTER GRADES

Techniques Revealed

College Survival Guide

Dr. Lane Bernard

Published by MaxGrades
A Division of Advanced Solutions, Inc.
Burke, Virginia 22009

Oh Yes You Can
GET BETTER GRADES

Copyright © 2004 by Advanced Solutions, Inc.

ALL RIGHTS RESERVED
Advanced Solutions, Inc.
Annandale, Virginia 22003

No part of this book may be reproduced or transmitted in any form or by any means - graphic, electronic or mechanical, including photocopying, recording, taping, or by any information storage retrieval system - without the expressed written permission of the publisher.

ISBN: 0-9745949-0-3

Printed in the United States of America

Words From The Readers

"Impressive and well done." - Leon L.

"I wish I had read this book before I went to college." - Jane S.

"I'm using it in my work and I'm not even in college yet." - Jim L.

"An expert shares his secrets – how powerful...how insightful! Lane's book is extremely motivating to those who require a push to see them through the quagmire of academia." - Laurie Ann L.

"WOW – this book says it all! How many students really realize that college is a stepping-stone for life? This book should be a great help for students as they journey through their college years towards a successful life!" - Lynda R.

"This book is a real life story. It shows that hard work coupled with believing in yourself can help you accomplish your goals in life! The book is very enjoyable and easy to read." Mary S.

"Thanks for a great book! As a returning adult student (a senior at a local University), I can remember – even back in high school – kids that sat in the front of the class seemed to get the best grades. I was not one of them. Today, I always get to class early just to get that ONE front and center seat. Sitting in the 'one special seat' is not the easy way out, but I believe it puts me in the best position to learn, and to get the best grades. Since returning to the University, I have used this as well as many of the other techniques in this book. I believe they have helped me improve my grades as, since returning to the University, I have never missed the dean's list and my GPA is nearly 4.0. Thank you for sharing these invaluable tools!" Barry D.

Many Thanks

A great deal of thanks goes to my wife, MJ, for her encouragement in writing this book, her many hours of reading it and discussing all that is in it. I owe her a great debt not just for helping me write about it, but also for living through it.

I owe a special thanks to LMR who had the painful job of carefully reading each version – not once, but three times - as the book developed. I can only admire and express my deep appreciation for her dedication and attention to detail.

I would also like to thank the numerous people who reviewed the many versions of the manuscript as the book grew and took shape. Including authors of other books, college graduates in many disciplines including English, Journalism, Teaching, Engineering, and Medicine, college and high school students, parents of college and high school students, college and high school teachers, college and high school counselors, high school principles, and those who just have an interest in helping students. Each was asked to provide a critical review. Their efforts are truly appreciated and their comments were a significant help to me in the development of the book. I also appreciated their encouragement and, in particular, the knowledge that the contents of the book are something that every college student should read, not only as they enter college, but as they progress through it as well.

The information, material, ideas and techniques contained in this book are provided for information only. Neither the publisher nor the author make any claim whatsoever as to their utility by others or the advisability of others to utilize or attempt to utilize them. If the reader, or anyone else, decides to try to apply any of the information, material, ideas or techniques contained herein, they assume full responsibility for those actions and the results that those actions may bring. Neither the publisher nor the author shall be responsible for the results of any application or attempted application of any of the ideas or techniques or any derivative or modification of an idea or technique herein described.

TABLE OF CONTENTS

FOREWORD xi

PREFACE xvii

INTRODUCTION 1

   1. Background 1

   2. Why The Techniques? 7

   3. Are Good Grades Worth The Extra Effort? 11

   4. College – A Real Training Ground 17

TECHNIQUES TO HELP IMPROVE YOUR GRADES 21

   1. Tried And True – With Some New Twists 25

      a. Attend Class 25
      b. Take Notes In Shorthand 26
      c. Develop Good Study Habits 27
      d. Check And Double Check 30
      e. Stay Awake – New Material Coming 31

2. Simple But Profound 33

    a. The Only Seat To Sit In 33
    b. Look Sharp, Feel Sharp & Be Sharp 34
    c. Be Active In Class 40
    d. Use All Of Your Senses 42
    e. Pace Yourself 43
    f. Plan Your Study Time 45
    g. Never Give Up 51
    h. Reward Yourself 51

3. Talk To Your Professors – With Purpose,
Understanding And Tact 55

    a. Two Minutes To An A 57
    b. In The Nick Of Time 58
    c. A Semester Of Hard Work 60
    d. Hostile Professor 63
    e. Betting On The Final 68
    f. A Flat No 68
    g. Beyond Hope 70

4. Power For Life - Employing Your Subconscious 73

APPLYING THE TECHNIQUES 79

EPILOG 83

FOREWORD

This book is about ways to improve one's grades. It differs considerably from other books on the subject. Where other books describe the conventional approaches like how to study, this book takes you to the next level. It describes techniques that will enable you to greatly improve your performance. Techniques that can add real power to your efforts. Power that will last a lifetime.

You might look at this book in the following way. Inside of most students there is a straight A student that would like to come out. The techniques in this book can help that straight A student emerge. Further, straight A students would like to maintain a straight A level of performance even in difficult times. These techniques can help with this as well.

This book is meant to help college students and prospective college students. Students generally have very little time to themselves and what time they do have they are not inclined to use to learn more. They are more likely to use any available spare time to get away from studying. Thus, a book of most any significant size that smacks of non-required studying, no matter how good it may sound, will probably end up in the read later (or more likely, never read) pile. Students need a book that can be read quickly and that offers immediate help.

With that in mind, this book was designed to enable the reader to quickly get to the point of improving one's grades. One can get initial ideas on things to implement in just 15 minutes. In 2 hours one can learn a significant part of what is in here. When put into practice that should be sufficient to make a real impact on one's grades. And it gets better. For example, consider the subconscious. While learning how to use the subconscious may take some time, it probably has the longest-term impact. Probably for the rest of one's life. However, while in school, think of it as working with two brains rather than one. If you were in an airplane race that you really wanted to win, would you settle for a one-engine model if a two-engine model was available?

While this book does provide techniques to help one improve his or her grades, it also provides many other benefits. For some, the notion that there are numerous factors that impact one's grade in a class, many of which are under their control, will be so profound that it will be like the dawning of a new day - perhaps a new era in their lives – a true wakeup call which will enable them to reach heights they never before realized they could attain. They will come to realize that grades are important and that good grades can help them achieve the goals they hope to attain in life.

For others, just knowing that help is available will be enough to encourage them to try college or to try to get better grades. Still others will expand on the techniques or even develop new ones. Still others will begin to understand that they are solving problems and will work to develop this skill. After all, to a significant degree, problem solving is what college is all about. And, to a significant degree, that's what companies hire employees to do. Further, that's what life demands that we do. How well we do in life depends a great deal on how well we develop our problem solving skills.

In addition, there are those who will realize that the material presented in this book is an introduction to a highly valued talent that goes by the popular name of "thinking out of the box". They will realize that college provides them with a unique opportunity to develop and refine such skills. And finally, they will come to realize that while what might be referred to as "book learning" is a significant part of the reason one goes to college, there is so much more to be gained while one is there. College is truly a unique training ground and most of us only get to go through it once. Thus, we should make the most of the opportunity while it is available to us.

For those who try to help students - parents, relatives, friends of students, counselors, etc. - this book can serve as a reference from which they can draw ideas to help those they care about. This book will help them understand some of the challenges faced by students and provide a basis for discussing solutions with them.

Some parents, and some students as well, might wonder why help might be needed – after all, the students did graduate from high school. While not wishing to add to the concerns of students entering college, four points are worth considering here; first year success rates, it's a big world out there, high school person – college number and there is a time for everyone. Let's look at these one at a time.

The latest figures indicate that up to 60 percent of those entering college for the first time don't make it through the first year. While hopefully the college that you have selected has a better record, you can be assured that a substantial number of those entering college with you will not make it. One might get the idea from this that it's not getting into college that's the problem – the real problem is staying there and of course graduating in a timely fashion.

It certainly is a big world out there. Depending on the size of the high school you attended and the college that you are going to, there maybe ten to one hundred times more students. A large high school might have 4,000 students whereas a large university might have 40,000. In most colleges the activities, both academic and non-academic, are non-stop. It's easy to get confused, or just plain lost, in that kind of environment. In general, organizational and efficiency skills play a bigger part in college than they did in high school.

Usually in high school one has the sense of who he or she is. Generally, we tend to go to school in the neighborhood we grew up in, go to school with students we have known much of our lives, take classes from teachers who know us or have at least heard of us and follow patterns guided by our parents and teachers. When we go off to college much of this gets left behind. We get a whole new set of friends, each of which we must calibrate. We also get a set of professors, which may differ from out teachers in high school in several respects. At least some who at least appear like they don't care about us and don't want to get to know us.

In general, everyone has their time – their time of need. The most intelligent person I ever knew, and he was well beyond anyone else I have ever met, had difficulty in his Doctorial program at MIT. He was unable to pass one of the language requirements on his own and he wouldn't ask for help. He never received his degree. Don't let this happen to you.

PREFACE

This book contains a set of techniques that I developed to help improve my grades in college. If you are trying to improve your grades in school, help someone else improve their grades or are just fascinated by the subject, this book can help. I have helped others learn the techniques. You too can learn them. You can also help someone else learn how to utilize them. Some of the techniques are very simple and can be applied immediately. An example of this is knowing which seat in a classroom to sit in. Others like planning and using the subconscious may take a little time to master.

Do the techniques work? You be the judge. Here are the facts. I did not like, was not interested in and did not understand the purpose of grade school. Thus, it is understandable that I graduated from high school with a dismal academic record. The real world soon showed me that I had missed the boat. Personnel directors at a number of large companies told me to go to college and get the best grades I could – that's like telling a drowning man he should get more air in his lungs. However, it was sound advice and I started to work on it. It took me **five** years to recover from the mistakes I made in high school, to establish a source of funding and to get into college.

When I finally arrived, I discovered that I was way behind the other students. If I was going to succeed I needed an advantage, so I began to develop the techniques described in this book. In no way do I want to imply that the techniques are a replacement for studying – they are not. They just gave me the advantage that I was looking for. And, in a bit less than four years, I graduated with my Bachelor's degree in Electrical Engineering from the University of Florida (UF), one of the major universities in the country, with an honor point average of 3.94 out of 4.0. Most people think that's pretty good. Those who knew me before I got into college thought it was a miracle.

I continued to refine and apply the techniques in my graduate work. I was able to complete my Master's degree in Engineering in nine months at the same university. However, a more significant test came when I went to get a Doctorate in Systems Engineering from the Massachusetts Institute of Technology (MIT), one of the best engineering universities in the country. When I arrived my advisor told me that I should expect to be there a minimum of four and one-half years. I finished in 27 months.

While I was in school, I wondered if by applying these techniques, I was having a negative impact on learning. I've come to realize that the impact is quite the opposite when the techniques are properly applied. I didn't develop the techniques to replace studying. I developed and used them to supplement my normal studying and I did study hard. However, I believe that techniques like *Time Slices* helped me to make sure that I applied the right amount of study time to each subject while techniques like *Which Seat To Sit In* kept me primed to get the most out of each class.

Additional good news. I have found the development and refinement of the techniques, as well as the techniques themselves, to be helpful in my career as well as in my personal life. I still use *Time Slices* at work and at home to ensure that I move each task along at a rate to complete it within its allotted time. Further, I believe that the semester by semester effort to try to reach a goal, in this case the best grades one can, sharpens one's problem solving capabilities - and life and career are certainly full of problems.

If you have any doubts about the importance of grades or the potential impact on your life, you might want to read the section, "OK, So Who Really Cares About Your Grades?" in the Epilog before you continue.

INTRODUCTION

1. Background

From the preface you know that I have a Doctorate in Engineering from one of the foremost engineering universities in the country. This may lead you to assume that I was one of those wiz kids in high school to whom high academic achievement seems to come so easy. Nothing could be further from the truth. I actually missed the boat in high school. I neither liked, was interested in nor understood the purpose of grade school – and I had the grades to prove it. I grew up four blocks from the Gulf of Mexico. I was an accomplished, award-winning sailor and water-skier before I was a teenager. I had various part-time jobs since the sixth grade. I bought my first car when I was fifteen. It wasn't much. I often had to push it to start it, but it was mine and it got me around. I was having a great time. On the other hand, had you asked my teachers and fellow students about my potential to succeed in the adult world – well you can probably guess what they would have told you – after they stopped laughing. Had you asked them for their thoughts on the possibility of me becoming a systems engineer, being paid by the government to evaluate the suitability of major weapons systems for utilization by our military, heading up technical research and development projects designed to push the state of the art forward or heading up major problem solving teams – well you probably wouldn't have wanted to wait around for the laughter to die down.

OK, so what happened? When I graduated from high school I had the good fortune to go to work for the Clerk of the Circuit Court in the county where I lived. For me it was a fun job. It paid very little, but I didn't need much. I spent much of the day behind a counter serving the public that came to the office for help. I met well-educated bankers, lawyers and people from the real estate community. For more than a year these people leaned on me to do something with my life. After a while I began to get the message. These people dressed well, were articulate, and drove really 'neat' cars. I also saw some of the exquisite houses they lived in. I wondered how I was going to do it. As I looked around I noticed that the price tags on things like new cars, travel, exotic vacations, a house of my own, and a family were much higher than I ever imagined. I began to realize the importance of having a job that would permit me to earn the money I would need to acquire these things in which I was developing an interest. I also realized that by goofing off in high school I had missed my opportunity to begin to acquire the skills needed to obtain such a job. The realizations of adulthood began to sink in.

Yes, they got my attention, but it was like putting a high performance engine in a cheap car. It's not going to do much unless one upgrades the rest of the car. I once noted that a car manufacturer offered a rather small and inexpensive car with an engine larger than the biggest one available in its sports car. My understanding was that it indeed had great acceleration and performance as long as one didn't need to turn or brake. One would also wonder if the car's drive train would be able to handle the torque available from the bigger engine. I began to realize that my new ideas about life were starting to upgrade my engine and that I needed to develop a game plan to upgrade the rest of me to prepare for the turns, and the stops and starts that are part of the real world.

So that's what got me started. However, this book can only cover one aspect of the transformation process. Details of my game plan, its execution and the resulting impact on my career, as well as the rest of my life, are contained in an upcoming book[1]. This current book is dedicated to describing the techniques I developed to get the best grades I was capable of obtaining in college. While utilizing the techniques certainly had a significant impact on my grades, I believe that the process of developing and refining them probably had as much of a positive impact on my life as did the classes I took in college. This is an important aspect and I will discuss it more fully later.

Just to clear the air, let me state very clearly what you will not get out of this book. In this book you will not learn how to cheat, bribe, butter up the teacher or any other objectionable, immoral, degrading or character destroying concepts. The techniques in this book are not aimed at providing an easy way out which once exposed can destroy all of what one is trying to achieve. I believe if you have to utilize undesirable techniques to get through college, you will have to utilize them to get through the rest of life. Not a very desirable aspect in my "personal" book. For the record, in all of the courses that I have taken and in all of the discussions that I have had about the techniques presented in this book, I never once met a professor or anyone else for that matter who had any negative thoughts about the techniques or the application of them to improve one's grades. Clever yes.

---

[1] "Oh Yes You Can – Change Your Life", in process

Objectionable no. Thus, what you will find in this book are professional level techniques that can help you in college as well as later in life. I say college because I have no experience in trying to utilize them in high school, trade school or any other type of schooling nor do I know of anyone who has tried to use them in any school system other than college. On the other hand, I don't know that they would not work in other situations.

Another question that needs to be addressed is, "Does applying the techniques reduce one's learning of the subject material?" No, I believe the opposite is true! The techniques are not designed to replace studying, but to *enhance* it. I believe that anyone who will go to the trouble to learn and apply, and perhaps refine, the techniques is a serious student. I believe that when a student, who is actually trying, begins to get better grades, that their image of themselves begins to improve. I believe that they begin to believe that they really can do better and they actually begin to work harder. I think techniques like *Look Sharp, Feel Sharp & Be Sharp* give a student a renewed feeling about themselves so long as they don't let their peers influence them otherwise – and why should they? Are their peers going to hire them? Not likely.

I also believe that talking to one's professors, in an adult sort of way, gives one a better appreciation of what the course is all about. Not that I would suggest for one moment that having a better appreciation will make it any easier or fairer. But whoever said college – or life for that matter – is fair. I do however suggest that learning to cope with the trials and tribulations of college will help one cope with those that will occur later in life. I also suggest that college is a great training ground. In a four-year degree program – eight semesters of say five courses each – one gets forty chances to get it right. Where else in life does one get such an opportunity? This is another very important concept – for parents as well as students – and I will discuss it more fully a bit later.

2. Why The Techniques?

This might seem somewhat obvious, but it's not and the reasons for it are instructive. The timeframe of my recovery plan for going to college was not short. It took me five years just to get into college. When I arrived I immediately found that most of the other students in my classes were fresh out of high school and far ahead of me academically. What I remembered of my high school classes – well let's just say it wasn't worth remembering. In addition, I had not taken any college preparatory classes. Further, I had chosen electrical engineering as my college major and the counseling I received from talking to personnel managers from some of the larger companies in the country was to get the best grades in college that I could. They impressed upon me that my future life was going to depend heavily on what grade point average I graduated with and where I stood in my class. I didn't realize how right they were, but since each one of them gave me the same story and laid it on me pretty heavy, I got the idea.

Now you can see the dilemma here. I had finally made it into college only to discover that the competition was far better prepared than I and I was supposed to somehow figure out how to rise to the top. Heck, I didn't even know how to study. At this point I was not so concerned about being an A student – I was just trying to survive. I had spent a lot of effort just to get into college. Having developed a second chance to define my life, I was not about to give up without a struggle. There is a fine line here, but to me it made a big difference. Trying to reach an objective in life can be a terrific motivator, but it's a self-made motivator. Survival is a built-in motivator and few human drives are stronger.

So what did I do? Well I got to thinking about how grades come about. Now that wasn't easy since for the most part I had not, in the past, given much thought to grades. So I started studying the subject of "how do grades come about?" I knew that learning the subject generally led to good grades on the exams, which generally led to a good grade for the class. However, I had seen things that gave rise to the idea that there were other elements at work in the grading process. I wondered how much of a factor they were and if there were other elements at work that I had not yet seen. Today, this is often referred to as "thinking out of the box". I observed, I did research and I experimented and experimented and experimented. Over time I found a number of techniques that helped. How much did they help? Well you decide. You know where I started. By carrying up to 21 hours a semester I was able to graduate with my degree in Electrical Engineering in about 3 ½ years with a honor point average of 3.94 out of 4.0. Most people think that's pretty good. If you knew me before I got into college, you would think it was a miracle. A further benefit is that I not only continued to use these techniques effectively during my college days (e.g., a Doctorate in Systems Engineering in 27 months), but I have also found them to be extremely helpful in my career as well as in my personal life.

Just so that you won't think that I took an easy path through all of this, let me give you the rest of the story – well at least a bit of it. I didn't do all of this alone. I got married during my first semester in college. A bit over a year later we had our first child. The longer I stayed in college the more children we had. Our fourth and last one was born while I was working on my doctorial thesis. Now all of this tends to increase the stress level, reduce the amount of time available for studying and increase the need for funds. It is clearly evident that, being late getting into college and with a family to look after, time was not on my side. Thus, I was inclined to accelerate my programs. Techniques that could increase my efficiency, quicken my pace and even perhaps help improve my grades were certainly of interest.

Finally, there was the unexpected benefit – already noted – of actually developing one's problem solving skills – certainly a skill set worth having while in school as well as after graduation; in one's business life as well as one's personal life.

3. Are Good Grades Worth The Extra Effort?

Few students whiz through college getting topnotch grades without putting some work into it. Further, we are not discussing techniques for students who are happy with whatever grades they get by just cramming for tests or even those who are just happy with whatever they get. We are talking about serious, hardworking students who wish to really make a change in their grades – say from mediocre to outstanding. So, before discussing how to improve one's grades we should address the question, "Why Try To Improve One's Grades?" That is, do they really matter? Are they worth the effort? The answer is a big YES! The reasons are rather simple, straightforward and a bit obvious – at least after the fact. Let me explain.

I did come to realize, after I graduated from high school, that one's high school grades play a role in one's acceptance to a university. However, up until I talked to a number of college recruiters right before I entered college, I really had no idea of the true impact of grades, good and bad, on one's life. Even then, until I actually saw it in action, I did not fully appreciate the real impact. Most of my understanding came from direct experience, as I will describe in the following.

The first indication came as I was starting my second semester in college. Because I had received good grades in my first semester, an organization, through the college, paid for my second semester's tuition. I kept my grades up and never had to pay tuition again – not for my Bachelor's degree (or for my Master's or Doctorate degrees, as I will discuss in a moment). We are talking big bucks here – just for grades.

Another early indication came during my junior year. Generally, to earn a Bachelor's degree one must take certain required courses. However, I found out that if one had sufficiently high grades, these rules did not strictly apply. That is, I was able to substitute courses I wanted to take for courses others were required to take. In effect, I got to steer the ship.

Another important indication came during my senior year in college. In large part, because of my undergraduate honor point average, I was offered a Ford Foundation grant for my Master's and Doctorial work. I still remember well the day I received the offer. Probably because up to that point I had not given any thought to continuing my education beyond the Bachelor level. I was concentrating so hard on finishing my Bachelor's degree and getting a job that the offer caught me completely by surprise. In truth, at that point in time, I had no idea how a Master's or Doctorate degree might impact my career. However, I did rather quickly realize that this would give me an additional opportunity to change my life and, although as it turned out I only elected to take advantage of the Master's part of the offer, it set up events that created additional opportunities a short time later.

I received another indication about the value of grades as I began interviewing for a job. A number of companies talked to me about working in their Research and Development (R&D) departments or organizations. Now this may or may not be an area of interest to you, but it was the area in which I wanted to work. When I asked why they were asking me about working in R&D they told me it was because of my good grades. One company told me that they didn't invite anyone to work in R&D unless the person had a grade point average of at least 3.5 out of 4. Now, like I said, R&D may not be of interest to you, but the message is, with rare exception, that the higher your grades are the more influence you will have on the jobs you will be offered after you graduate. In fact, as I will explain in a moment, the more opportunities you will have – period.

Each time I graduated and went to work, I was fortunate to be asked by that company to accompany one of their college recruiters back to the college I had just graduated from to help with the recruiting. These efforts provided considerable insight into how grades impact our lives. Getting to see prospective college graduates from the company's point of view was very enlightening – in fact it was more of a revelation. It gives one a totally different prospective on grades as well as other characteristics that students live with every day and probably have no idea that the recruiters are looking at them.

On a typical college-recruiting day for the first company, the real recruiter and I would each interview about 24 candidates – spending about 20 minutes with each candidate. The time went something like this; first 5 minutes - make candidate feel at ease, next 5 minutes - find out what candidate wanted to do, next 5 minutes - decide if company would be interested in candidate and last 5 minutes - tell candidate of decision, but put in such a way that candidate felt good about it and of course would keep buying the company's products – for example, cars. Then, at night, we would review the applications and our notes from the interview for each candidate and discuss what each of them might be able to do for our company. I wouldn't say that grades were the only item that drove us to choose one candidate over another, but they certainly played a dominant role. When one realizes that a college recruiter has a very limited amount of time to learn about each candidate and they often see a large number in a short period of time, one can see that certain significant things might stick out in a recruiter's mind. For sure, very good grades and bad grades both stick in a recruiter's mind. Further, when one realizes that a recruiter is not making the final decisions on the candidates, but only getting his thoughts together as to how to present his findings to the managers in the company, one comes to understand that there has to be something significant about a candidate for the recruiter to be able to "sell" the candidate to the company's managers who are then going to decide whether or not to make an offer to the candidate. Often this "selling of a candidate" is not an easy task. In my experience, something about the candidate had to lead us to believe we could make the case to the company's managers that the candidate could help our company. Couple this with the age-old adage "You only get one chance to make a first impression" and then look to see what is the first thing a recruiter sees about a candidate.

Often a recruiter will have some background on each of the candidates he or she is scheduled to see during the day - often a form that the candidate fills out which includes the candidate's grade point average, the type of work that the candidate is seeking and the college program that the candidate is in. The recruiter uses this information to prepare to talk with the candidate.

Now let's put all of this together. The recruiter is looking for things he or she can use to "sell" the candidate to the company's managers. The recruiter's time with each candidate is very short and often the first thing he or she sees is the candidate's grades. If the recruiter is looking for top candidates the fate of those with poor grades is virtually set before the candidate gets to talk with the recruiter. Further, there are often the breakpoints. For example, as I already noted, to get into the research area of one of the companies I worked for, a candidate had to have at least a 3.5 out of 4.0 average. No exceptions. Thus, students with less than a 3.5 average never heard about the opportunities in research.

On the other end of the spectrum, the recruiter may be looking for grunts – candidates to do the grunt work. In that case, candidates with average or mediocre grades might be "saleable" to the managers because they can be hired at a much lower salary. My experience as a recruiter indicated that there were few candidates who spent their time in college aiming to be a grunt.

As my career progressed I became a manager. Then I got to see the rest of the recruiting process. Now I had to make the final decision on making or not making offers. Still, for candidates that are graduating from college, especially those with no real experience, one has precious little to indicate how well the candidate will do in industry. What the candidate did in college may still be the best indicator of what he or she is capable of doing in industry. I believe that most managers would prefer to have better indicators, but they are forced to resign themselves with the realities of life. I also came to realize that the interview with the recruiter is just the first step and that many managers really do rely on the recruiter's observations in whether or not to invite the candidate to come to the company for the next phase of interviews. Thus, a candidate with average or poor grades may never even see the more interesting jobs.

One's accomplishments in college, especially one's honor point average, will follow one throughout his or her career. Although more evident in the early years, grades continue to be a factor even in later years, especially in considerations associated with promotions or new jobs. In my own case, the company I went to work for after I received my Master's degree knew that I had a fellowship to continue with my Doctorate. They knew that I was leaving the University because I did not relate to the coursework the University had to offer at the Doctorial level in electrical engineering. This was very fortuitous because, although I did not realize it at the time, my concept of what I wanted to do was changing. I began to realize that as much as I liked electrical engineering it was only a part of the engineering process – albeit a central part. Thus, when I joined the first company, its upper management asked if I was still interested in obtaining a Doctorate. A year later they offered me a company fellowship to continue my education. Obtaining a Doctorate certainly has had a profound impact on my career, which has had a positive effect on my personal life.

4. College – A Real Training Ground

College certainly is a training ground. However, I believe that maybe only half of what college has to offer is in the "books" – that is, the traditional learning for which most people believe they are going to college to obtain. To me college is a training ground in a much broader sense. To a degree it provides a transition from adolescence to adulthood. As such it provides a great opportunity to *learn how to operate the controls of life*. Think about it this way - life is not fair and neither is college. However, if we look at college as a training ground for life, instead of just more book learning, then the opportunity that I am talking about becomes more clear. That is, for example, the restrictive rules and regulations, the professors who appear to be disinterested and perhaps unfair, and the uninteresting classes we must take are just part of life.

A typical student takes about five classes a semester for eight semesters to obtain a Bachelor's degree. That's 40 classes. Forty opportunities to try to understand how that portion of college operates. I'm sure you will agree that there are many more when combined with housing, eating, entertainment, money and the many other problems with which a student must cope. Learning to deal with these problems is a good deal of what college is all about. As unpopular as the thought might be to some, if you look to others to solve these problems for you – like your parents – you will miss one of the greatest transition opportunities the world has to offer. Oh yes, it's hard, it's inconvenient, it takes time and your peers may even laugh at you. On the other hand, the next step is real life and you'll meet the same problems there.

Make a mistake in college and you might get laughed at or a lesser grade than you had hoped to get. A mistake in real life is a whole different story. If you were going to run an important race with lots of competition you know you would go and practice. Well you are getting ready to run a big race – it's called life – and college provides an opportunity to practice.

And grades are an important part of this opportunity. Studying is certainly fundamental to getting good grades. Generally, hard work is a fundamental element in advancing in your career, in making money and in achieving a happy life. At least on a level playing field. However, whoever said the playing field was level? As already noted, life – and college is part of life – is not fair. Most people that I met in my college days assumed that the grading process was pretty simple – one took courses, studied, took exams, received grades and, of course, grumbled when the grades were not what one was hoping to receive. Because of my already mentioned situation, I looked a bit deeper into the process, trying to understand the rules to see if I could improve my chances of success. For the most part what I found was that there were no well-defined rules. Perhaps each class had its own set. Perhaps there were no rules at all. At best they were obscure, poorly defined and conflicting. So how does one get ready to compete in a world with an obscure, poorly defined and conflicting set of rules? The more I asked the question the more it sounded like real life.

OK, so what's wrong with trying to solve these problems in a positive way? Since college is a training ground why don't we practice by trying to get better grades – in a positive way of course. It certainly has a nice *ring* to it. And, there is a real reward for success – better grades. In addition, there is almost no penalty for failure – one simply gets the grades one would have received had they not tried. Further, it has a somewhat built in measure of success. That is, at the end of the semester or term one can take stock – I tried this technique or combination of techniques – are my grades for the different classes better than what I would have expected if I did not apply the techniques? But don't stop there; reflect on how you can improve on the techniques in the next term.

In my own case I felt the need to do more. I decided to try to add some excitement to the process by making a game out of it. Each course represented a different set of problems. Each required its own solution. The object of the game was to see how many of these problems I could solve each semester. The measure of effectiveness was the number of A's that I got each semester. I began to really enjoy the game. I not only learned the class material, I began to learn about solving problems. I have already told you of the results in college. The results after college are contained in the already referenced upcoming book.

TECHNIQUES TO HELP IMPROVE YOUR GRADES

The techniques are divided into four categories; Tried And True, Simple But Profound, Talk To Your Professors and Power For Life. Most of those in the first category, Tried And True, you have heard before – probably ever since you entered grade school. I'll not bore you with that which you already know, but I did find a few interesting things in my study of the subject which I will pass along to you. I do not, in any way, mean to downplay the importance of the techniques in the first category. However, I expect that every student knows and applies them to some degree. On the other hand, I want to make sure that you are already applying all that you know. This book should enable you to build on that.

The second category of techniques, Simple But Profound, is a bit more interesting. Generally, although they are easy to apply, they do take some effort. The simplest sounding of all the techniques – *The Only Seat To Sit In* – is, as are all of the techniques, for the serious student as it will make unimagined demands on you. However, I believe the serious student will realize those are good demands and will want to sit in that seat. Other techniques in the group like *Look Sharp, Feel Sharp & Be Sharp* and *Be Active In Class* help one cope with the demands of sitting in the Only Seat and complement the rewards of sitting there.

The third category – Talk To Your Professors – With Purpose, Understanding And Tact may only seem like a group of examples and to a degree they are, as I have not yet found an adequate way to categorize the different reasons to talk to one's professors and the best approach to use in each of those cases. Further, as strongly as I believe that one should talk to his or her professors, the concept comes with a warning label. While the application of most of the other techniques do not, for the most part, present an opportunity for a down side effect, this one does. Professors are human and they see students looking for an easy grade in most every class. I certainly saw it when I taught. So when you ask a professor for help, be careful what you ask for and be ready to do some serious work. I believe this category, more than any of the others, is only for the serious student, because if you ask for help and you don't follow the professor's advice you may appear to be an insincere student looking for an easy grade. This may place you in an even deeper hole than you were originally in – a hole from which you may never emerge – at least not in that class. And just like success can lead to success, disaster can lead to disaster.

The last category – Power For Life – is a discussion on Employing Your Subconscious. This technique cannot be applied directly and it takes time to learn. Communicating with one's subconscious is a bit of a challenge. Further, measuring the impact of what the subconscious can do for you, even if you are able to communicate with it, may be even more of a challenge. However, this one technique can have a greater impact on one's life than all of the rest of the techniques combined. If developed and maintained, this technique can be applied in every aspect of one's life. It not only works in college, but at work and home as well.

The human brain is thought to have two distinct levels of activity, or as some would say, levels of processing – the conscious and the subconscious[2]. Most people work with the conscious part. The subconscious develops in response to the demands placed on it by the conscious part of the brain as well as the other parts of the body to which the subconscious is connected. When I started looking into this area, I had never encountered anyone who had tried to train or specifically develop the subconscious part. So I decided to try. Wow – having two brains working on an exam – why it almost felt like…well I could actually feel it. But, it needs to be coupled with study, because the processors are not much help if the information has never been put into the brain.

Let's be sure we are clear on what I was trying to do here. The subconscious does most of the processing. Without direction, the efforts of the subconscious seem to be somewhat random in nature. What I was seeking to do was to set up better communications with the subconscious so that I could better direct its efforts – like having it work on the exam problems *during* the exam, rather than later, like when I was trying to sleep that night.

---

[2] For a more complete discussion of the subject see "The Human Computer – Get The Most Out Of Yours," Writers Club Press, iUniverse, Inc. 2002

1. Tried And True – With Some New Twists

a. Attend Class

This might sound like something your high school teachers or advisors might have told you. Well when I went to high school they took attendance so they knew if we skipped class. When I went to college they didn't seem to care if we attended class or not, so we had to make our own decision. New freedom - new responsibilities. To attend or not? Well, as I said above, I was trying hard to earn the best grades I could. As part of this, I was studying what really went on in class – not just what the professor was trying to teach. It didn't take me long to discover that every once in a while something would come up that I would not have known had I not been in class. Sometimes these things had the potential to significantly impact one's grade. For example, from time to time professors may indicate what they consider important. They may indicate during class that certain things may (will?) be on the next test. Further, they may actually tell you what you should study for an upcoming test. Like "You should study chapter 4" or "We have covered everything in class that will be on the test."

Attending class and paying attention to what is happening around you, in addition to what the professor is saying, will often have other benefits. Listen to what other students ask and how the professor responds. What the professor's body language[3] might tell you and see what seems to please and displease him or her. And one more thing, I started this book by saying it was only for

---

[3] See Epilog for more on body language

serious students. It certainly makes it harder to convince a professor that you are a serious student and that you need his or her help and advice, as will be discussed later, when you didn't even bother to attend class.

b. Take Notes In Shorthand

In some classes knowing what is in the book or books is sufficient to do well. However, in many courses it is important to remember what the professor said and/or put on the black/white board during class. Which is more important - to listen or to take notes? Most of the time it is both. Most of the time it is not possible to do both – at least to listen while writing notes in longhand. A course in some form of shorthand would seem helpful. I didn't have time for that – I don't think any was offered. The students I know that knew shorthand – men and women – knew it before they entered college. I developed my own.

I felt that it was important to understand the points the professor was trying to make. Many times he would clear up things that were troubling me or give me insight into what he considered as important. The class notes were meant to reinforce these points. However, since I knew that my notes, for the most part, were incomplete and cryptic, I tried to organize and type (or rewrite) them later, as soon after class as I had the time. Sometimes I would develop them further as more information become available. Sometimes I would put the additional information in a different font so that I could tell what the professor said verses what else I had learned. I always tried to keep in mind that, for me, the intent of the notes was to enable me to more quickly study for the upcoming exams.

c. Develop Good Study Habits

This might seem obvious to you, but it was not to me. When I entered college I didn't know how to study. I never learned in high school. I understood the word, but not the real meaning. While I was in the Navy, I took several courses at a local university. As noted above, my plan was to get out of the Navy when my enlistment was up and attend college. Since I didn't know what college was all about, I took these courses to get a preliminary idea as to what I was getting myself into. Each of these courses had a book from which assignments were made.

Since I didn't know how to study I read both of the books over and over until I could see the pages in my mind. I wouldn't say that I had anything like a photographic memory. I just read the books until I could read from the pages in my mind – mentally flipping pages as I went forward. While it was helpful for taking exams – at least in these classes, it was not learning. For one thing it took forever. For another, the images didn't stay around very long. But more importantly, I couldn't use the information. I hadn't learned it. I didn't really understand it. I had only memorized it. If you can't use it, say in your next class, you really haven't learned it.

Thus, there is a huge difference between learning and memorizing. What you memorize you can generally repeat back. What you learn you can use. You can apply it to new situations. Most disciplines require some memorization, like the math tables. Some disciplines require a great deal of memorization, for example, medicine. In general, these are tools, not course material, and one is expected to apply the tools. Memorizing the book, i.e., the pages in the book, does not provide the subject matter in a way that it is easy to apply or expand upon.

However, there is not "A" way to study – there are many ways to study. The best way is the one that works for you. There are numerous books and websites on the subject. For example, do an Internet search on *how to study*. Far be it for a book of this size to try to expound on the subject of how to study. On the other hand, when you reach the point where you are able to use or expand upon what you have learned – for example in previous classes – in other classes or in life, you will know that you are headed in the right direction.

College requires one to read many books or parts thereof. When you read the book think about the points being made and the message it is trying to put forth. If it is your book, write in it. Make notes in the margins. Underline or highlight the important points. Try to do it in such a way that you can use those markings for a quick refresh of the material for exams – so you don't have to reread the whole book again. It seems there is always too little time when one is studying for exams. Marking in the book can be a real time saver. Another way is to make an annotated outline of the book. This may take more time, but it may save more in the end – at exam time. For some, it will make the material easier to remember. It's also a useful approach if the book does not belong to you.

Another good way to study is to teach the subject. Now it might be hard to get students to listen to a subject that you are just learning – but maybe not. It could be a class of your peers that are just a bit below your level. Or maybe the group is taking turns trying to teach each other. On the other hand, who said you need students. Go to the blackboard (or whiteboard) in an empty classroom and teach the chairs. It helps to write on a different media. It helps to get all of your senses engaged in the learning process. Be sure to talk as you write. And listen to what you are saying. Does it make sense? Does it flow?

While teaching others does have its advantages, like getting feedback and the ability to discuss the different topics, it does have its disadvantages, like the coordination required, the need to set a time and the need to have a more formal agenda. On the other hand, some of these aspects, which might be disadvantages to some, might actually help others by forcing them to do what they can't do on their own. However, one must guard against the group getting hung up on a particular topic or wandering off the subject altogether. If you are teaching the chairs you don't have to be well prepared. Just go to the blackboard and see if you can explain the subject – step by step. Where you can't, you need to apply additional work. Be sure to talk and listen while you write. If the chairs are not live enough for you, you might try your spouse, children or friend. Just because they may not understand the subject does not mean that it won't help you to learn the subject.

d. Check And Double Check

Get in the habit of checking and rechecking the work that you do. If you finish the test early, check your answers. If you have time, try working the problems in a different way – use approximations to see if your answer is close. Never leave an exam early unless you are sure that you have done the best you can. Sometimes just relaxing for a moment or a short diversion, like getting a drink of water, can reenergize the thinking process. This is true not only on exams, but homework as well. Watch for simple things that the professor might have done to throw the fast movers off the track. An example, from the technical world, is to change the units of some of the parameters in a problem – like having vehicle speed in kilometers per hour where all other measurement are in miles. Actually, the professor is not trying to trick the students as much as get them ready for the real world where this happens all of the time.

If you write a paper, reread it before you hand it in. Don't procrastinate, but use the time available. If you have a day or two after you finish writing a paper before you have to turn it in, then put it aside and reread it later. This is very helpful. You will see what you have written in a different light. Your ideas will generally be broader and more comprehensive. If appropriate, discuss the writing with others. Just discussing the subject often brings out holes in one's logic and causes new ideas to form.

e. Stay Awake – New Material Coming

This is an area where I messed-up big time and so have others I have known. This occurs when one knows the class subject before taking the course. All is well during the first part of the course. It's like attending a review and one figuratively falls asleep in the class. Then, when the new material gets presented, one does not even see it or is so busy that he or she can't adjust his or her schedule fast enough to accommodate it.

The way that it happened to me really hurt, but I'm not sure which hurt more the fact that I got a B in a course that I should have gotten an easy A or the fact that I was stupid enough to let it happen. As you read these techniques you will realize that I put a great deal of work into each course – well, almost every course – certainly not this one. I didn't have to – at least I didn't think I had to. It was my first course in digital computers; a subject for which I had developed a likening. I took it in the Spring semester. Over the Christmas holidays, prior to that semester, I had written a paper on the subject, which was submitted to and later published in a national engineering magazine.

The initial part of the course was very much along the lines of the paper I had written. My mind began to sleep – figuratively speaking of course – in that class. I got so used to not learning anything in that class that when new material was presented I failed to see it. One day I woke up – figuratively speaking again – and realized that I didn't know what was going on. However, it was too late. The class was too far ahead. There was not enough time to recover. I was lucky to get a B!

The problem with catching up was that I usually filled my days tighter and tighter as the semester moved along. Generally, there was never enough time as many classes assigned or suggested much more than I could get done. I tried to carry 21 hours a semester. So, as I will discuss later, I carefully planned not only my days, but the parts of each day and then the hours within each part as well. I called it *Time Slices*. I still use the technique today. However, what happened here was that I had arranged my workload such that I was truly maxed out. Since it appeared that I didn't have to study for the computer course, I didn't set any time aside for it. As the midpoint of the semester came and went and I was studying as hard as I could, and getting behind which always seemed to happen, I suddenly realized that I no longer knew what was going on in the computer course. I tried desperately to catch up, but the workload and the time remaining in the semester were against me and I just couldn't quite get there. I was angry with myself for being so stupid as to let an easy A get away when I had worked so hard to get others, but I assure you that I never made that mistake again.

2. Simple But Profound

a. The Only Seat To Sit In

When you are attending class or are in a classroom type environment, there is only one seat to sit in. It's in the front row right in front of the professor. Many people miss this simple rule. However, there are a number of reasons to follow it. One, you won't be distracted by what the other students are doing or not doing. If anyone can hear what the professor is saying it will be you and you will have the first shot at asking questions. Also, if the professor is writing on the board you will generally have the best view of what he or she is writing. However, there are some other less obvious reasons. One, as we will discuss later, is that you will want to get to know your professors. This seat will give you the best view of any unvoiced expressions, like facial expressions, made in response to things that go on in the classroom. This may give you clues as to what the professor expects, doesn't expect, likes, and dislikes. You will also have the best opportunity of anyone to get to him or her after class is over. In addition, sitting right in front of the professor will make you a better student. You will be more conscious of how you look. You will be inclined to be better prepared as you know the professor can see you and he or she may very well call upon you since you are directly in front of him or her. Obviously, in this seat, you need to look sharp, feel sharp and be sharp so let's have a look at these attributes.

b. Look Sharp, Feel Sharp & Be Sharp

This is not only a great saying; it's a recipe for success! It was given to me by the best boss that I have ever had. I describe more of what he taught me in an upcoming book[4], but let me tell you what I have taken these words to mean and how they might help you in school. However, let me preface this by saying that we can all find counter examples to the following, especially looking sharp. That is, we have all seen people that didn't look sharp, but they succeeded anyway. On the other hand, it's hard to imagine improving one's position by looking dumb, feeling dumb and being dumb. If that was your approach you wouldn't be reading this book.

Look Sharp

The saying "You only get one chance at a first impression" remains true in all walks of life. Further, overdressing for the situation is nearly as bad as under-dressing. Looking sharp means that you arrange your outward appearance to stand out in a positive way. It means adequate attention to all elements including hair, face and hands as well as clothing. It means clean. A bath will do, moderate amounts of perfume and deodorant for women and aftershave and deodorant for men are nice as long as they are not overdone or used as a cover-up.

For ideas on how to dress for your environment, look around for examples of what you would like other people to *see* you as. It maybe one person from

---

[4] ibid, "Oh Yes You Can – Change Your Life"

whom you get the ideas or it might be that you assemble them from different parts of a number of people. Also, there are a number of magazines that one can look at. Notice I said *look* at. From all of this information you can get at least an initial concept of how you would like people to *see* you. Then try gravitating toward that concept. As you improve, your ideas about what you would like to look like will change - hopefully for the better.

Like you, I have seen many people that looked outstanding. In many cases it was natural. However, one young lady I met stands out as an example for all of us. She was a real head turner. A real knockout. But, when you knew her as I did, you knew God had only given her average amenities to work with. What she had done was to take these average amenities and enhance them to the point of stunning attractiveness. She was not obsessed with looking beautiful. She simply wanted to look as good as she could. It made her feel good. It gave her confidence. And actually she spent very little time at it. However, she did spend a few minutes every day trying to figure out how to improve her appearance. All those I knew that met her considered it time well spent.

Actually, the message in the paragraph above goes well beyond *looking sharp*. It's about developing goals and taking incremental steps over a period of time to reach those goals. It's about persistence. It's about being aware of how one is doing and making adjustments to one's techniques to better reach those goals.

Try this test at every chance you get. When you walk by a mirror, window or other object that provides you a reflection of yourself, look at the reflection and ask, "Is this the person that I want others to see?" "Am I standing up straight, are my shoulders square and back, is my pot hanging out?" If this is not the person that you want others to *see* then decide what's wrong and pick something to work on. For example, if you observe that you are not standing up straight, your shoulders are not back and you are not pulling your stomach in, then try to do those things. And every time you walk past a reflective surface check to see that you are doing it the way you want others to see you. Soon it will become a habit, a habit that you will enjoy because you know that others are seeing you more like you want them to see you. It will make you feel good about yourself. You will even feel more confident.

Feel Sharp

How we feel about ourselves, as discussed above, has a good bit to do with how we feel in general. However, here we are talking about getting enough sleep, exercising and eating right. In general, the proper care and feeding of the physical body. The bottom line is, get a good night's sleep. We all require somewhat different hours of rest, but whatever yours is make sure you get it, especially before big events like tests. A few more hours of study might not make much difference on an exam, but taking an exam when you are falling asleep can greatly impact your grade in the wrong direction.

Eating right also has an impact. Eating right does not just mean eating the right food, it also means eating at appropriate times and in appropriate amounts. There are piles of books out there on eating so I will just provide a few guidelines I have learned over the years. Learn to eat what the body needs – not what *looks* good. Eat when your body needs it. Some people skip breakfast. That might mean 15 hours between meals. Now most people's level of concentration is highest in the morning and falls off toward evening. Thus, it never made sense to me to expect maximum productivity out of my body while I was starving it. I eat a moderate breakfast. Don't overeat and certainly don't eat a big meal right before a big exam. Much of the world stops to eat three times a day – yes, I know, much of the world never stops eating, but we are going to discount that at this point. Eating three times a day seems to fit our life style, but that doesn't mean it's the best way. To operate at top efficiency day and night for long periods of time it might make sense to eat smaller meals but to eat them more often. It's not an opinion, just food for thought.

Drinking water doesn't seem to fit in the eating category, but it plays an important role in the body's functions. The current recommendation seems to be eight 12-ounce glasses a day. Whatever it is be sure to follow it, but do not try to drink it all at once.

Another area that helps one feel good, or sharp as we are calling it, is exercise. The body needs a certain amount of activity to keep all of its parts functioning at a reasonable level. Again, there is so much material available on this subject that I will not add to it here. However, I will say that from time to time I did not get enough exercise. I didn't give it enough thought. I probably should have had at least a basic plan.

Well, it finally caught up with me. I was studying for both my doctorial exams and my regular classes. I was pushing pretty hard. One day when I was studying I just needed to stretch a bit. When I did, I pulled a muscle that goes down across the heart. I felt it pull. It didn't hurt – until the next day. I thought I was having a heart attack. The doctor's diagnosis was that I needed to get some exercise. No time, but when I graduated the next year I started a running program. As long as I ran the pain stayed away. When I stopped it came back. The pain helped me understand the body's need for exercise.

Exercise is not just for school; it is for your whole life. When I graduated I went to work. The stress was considerable, but I loved the environment. I did notice that the pressure had an effect on my body. My heart rate would go up as my turn to make a presentation arrived. On the other hand, I observed that when I was into my running program my heart rate rose very little as my turn arrived. When I didn't run for a while, it rose a great deal as my time neared. It didn't take me long to catch on. I ran for 14 years after which I started an aerobics program, which I still do. When I do it I feel great. When I don't I can tell the difference.

<u>Be Sharp</u>

You can look sharp and feel sharp, but if when you open your mouth nothing worthwhile comes out your chances of success drop dramatically. On the other hand, it is nearly as bad to not say anything when clearly you should be engaged in the activity, rather than giving the appearance of a well-dressed bump on a log. So do your homework. This is not a reference to classroom homework (although in most cases you will want to do that), but a reference to being ready for the event. For example, if you are attending a meeting tomorrow and it's a subject that you might be able to contribute on, study up on it before the meeting. If something is said that seems wrong, ask about it. On the other hand, don't fall into the trap of thinking that a few minutes of study makes one an expert. However, you can certainly say something like "Can you give us some examples of that?" or "How does that compare with Dr. Smith's briefing on that topic that he gave last week?" For sure, if you are the speaker, one of the speakers, or if you are giving a presentation, be sure you are ready.

One way to maintain sharpness is to follow the rule - Never Run At Idle. As does most of the material in this book, this applies equally well to work, school and to life. When you are at idle the world is moving ahead of you. You need to turn any available time into productive time. I'm not saying you should not do things like take vacations, have fun, rest or spend time talking with your friends. What I am saying is if you have some idle time don't waste it - try putting it to productive use. If there is nothing you can do for school or work, then look around for something to learn that will improve your position in school (for example, you might study how to study) or increase your worth at work (for example, how might the next version of Microsoft Office help you do your job better).

c. Be Active In Class

Like it or not, the classroom is a stage. And no, I'm not in any way suggesting that you become an actor, put on an act in class or that you use the stage to advance your own objectives. What I'm saying is that you should be aware that people, your peers as well as your professor, note what you do in class. They notice how you look, how you act, how you respond, how you talk and the merit of what you have to say. What you say can have a profound impact on them – both positive and negative, but these other factors can as well. Sometimes it's not what you say, but how you say it. Do you listen or just talk? Are you trying to get ready to meet the world, in particular the business world, or are you just trying to please your peers so as to remain part of the group?

By being active in class I mean participating – asking questions when you don't understand or aren't sure, volunteering answers to questions and, if the class is in a discussion format, giving your version of the idea or concept that is being discussed. Of course this means that you don't hog the floor and that you can accept others' points of view.

Being active in class also implies that you do the homework and that you do it on time. It will give you a chance to answer as well as ask questions about it. When reasonable, you might consider volunteering for extra class projects. You may actually remember more from these experiences than reading the textbook. However, I did say when reasonable, because there may be more to it than one would have expected up front. You will find more about this in the section on "Talk To Your Professors..."

Being active in class will give new meaning to studying. If you are going to be active in a positive way you will need to be sure that you have adequately addressed the material to be covered or discussed in class. Being active in class will also keep your attention on the material being presented in class even when you are not the one who is talking. You will want to be alert so that you can answer the next question or even ask one if the material being presented skips a beat or two. You will want to ask yourself if you really believe and understand the material being presented. Keeping that level of attention for a semester will certainly have a positive impact on your understanding of the material and, in general, the grade you receive for the class.

Being active in class helps in a number of significant ways. Probably most importantly, it helps make the material part of you. On the other end of the spectrum, you become part of the class. Assuming that you follow other ideas in this book, the professor as well as your classmates will begin to notice you as a serious student. Don't be afraid to speak out. Give them your version of the concept or idea. So what if you are wrong. Well, if it is a debatable issue then there are many points of view and you can have yours. If it is a question of fact and you are wrong, then admit it. Thank the person that brought it to your attention. After all, you are there to learn. If you knew it all you probably wouldn't be there. Since you are there, you should try not to leave with misconceptions. The only thing worse than not learning is learning things wrong. Also, don't forget to ask questions to clarify your understanding and to thank the people who help you to learn.

d. Use All Of Your Senses

You have five or maybe six senses. These are the means that information gets into your brain. Make sure that you are using all of them. Further, make sure that you are using them in combination as that tends to reinforce the information in your brain, which makes it easier to get the information back out.

The five normal senses are sight, hearing, smell, taste and touch. Some people seem to have a sixth sense. If you do, employ it as well. Obviously, in a classroom environment sight and hearing would seem to play the dominant role. But, don't forget the touch - in this case writing or taking notes. The combination of the three will make the subject easier to remember. Using different environments for studying can also help. For example, after you have studied at your desk for a while, move to working on a blackboard or whiteboard. Also, when combined with talking the ability to recall the subject is often even stronger. This is one reason that you want to participate in class. However, at other times, like when you are working at a whiteboard, talking about the subject while you work on it, even to yourself, but out loud, can help as well. In some courses taste and smell can be used as well; like in chemistry and cooking.

If you are giving a speech, give it to your dog or cat or try it from a blackboard. Go to an empty classroom or ask someone if you can give it to him or her. Going over it the night before, just prior to retiring may also help.

e. Pace Yourself

Getting a college education is usually a multiyear process – generally at least four years. You don't want to ever burnout - certainly not on your way to your objective. Once in college, it doesn't take long to realize that the semester's workload gets larger as the semester moves along. On the other hand, it is amazing how much one can do in an allotted amount of time when one puts his or her mind to it. I started out with the normal 15 to 16 hours a semester. However, I found I could do more; even in electrical engineering and with a wife and children. I moved the pace up to 21 hours a semester. I found this easier to cope with than 15 hours because I didn't need to find time to study - I was studying all of the time.

Well not quite all of the time. For one thing one needs to get an adequate amount of sleep – in my case back then, 8 hours a day – well most days. Like I said, getting an education is a long-term process and one needs to keep the body and the mind in good running order. It has been shown[5] that while one sleeps the subconscious works. It reorganizes the conscious mind, clears it of extraneous information and places important items in long-term storage. A significant amount of body rebuilding also occurs. You probably have noticed the effect of this when you got up this morning. Most people

---

[5] ibid, "The Human Computer"

feel tired when the day ends. When they wakeup, after a good night's sleep, they usually feel good and clear-headed. When they do, the subconscious along with the other parts of the human control system and the elements the subconscious controls have done their job[6].

Further, studies have shown that having set sleep and wakeup times, especially on school days, is beneficial. Now you may find this difficult to accomplish – I did – but it's a good thing to keep in mind. In some of my classes there were team efforts. Sometimes a team didn't do its work until the night before the results were due. In those cases I had to stay up all night, much of the time talking to other members of the team, to get the work done as we were not permitted to do the project alone - the rest of the time was needed to type the report on the project.

So one must do what one must do, which means that sometimes one might not get an adequate amount of sleep. This should not be a problem in the short-term, but the message is *don't expect to do it in the long-term and win any medals*. Another thing, don't study all night before a big test. Yes, you feel the need to study. But chances are that the reduced capability from the sleep you lose will more than offset the gain in knowledge that you make from studying all night. You need to make a tradeoff and the better tradeoff is to study before time and over a period of time.

The bottom line here is that each of us must find the pace that will allow us to reach our objectives over the long haul. If the undertakings are too demanding and the

---

[6] "The Human Control System", in process

pace is too fast one might not be able to get everything done. This can have disastrous results especially toward the end of the semester. Too slow and that may very well lead to too much time on one's hands – which can be just as bad.

f. Plan Your Study Time

In college, as in life, there is usually more to get done than time allows. Getting more done each day in college can lead to better grades. Being more efficient may help reduce the stress. It might help one to reach certain objectives like taking some additional courses that are of particular interest or even to graduate early. Being organized will help to make sure that one doesn't miss the due date for a term report and that one does allocate sufficient time to study for each exam. This technique is equally applicable to getting things done after college, both on the job and in one's personal life. There are three parts to this technique. We'll call them <u>Calendar</u>, <u>Approach</u> and <u>Plan</u>.

Let's talk about the <u>calendar</u> first. You should maintain a calendar of important events like the date and time of upcoming exams and due dates for term papers. That is pretty straightforward. A semester is typically 16 to 18 weeks counting finals. You can make a calendar to cover that many weeks on no more than two sheets of paper (I currently maintain a three month calendar on one sheet). Carry this or something equivalent around with you and update it as new tasks are assigned and new plans are made. If this is inconvenient then keep it in a place where you are sure to review it every day to make sure that it stays up-to-date. The only thing worse than not keeping a calendar is keeping one that is incomplete and/or out-of-date.

Next you need to determine how you are going to get each of the events done that you have entered on the calendar. That is, you need to develop an <u>approach</u> to insure that you accomplish each one in a timely, efficient and adequate manner. To do this you need to determine how you are going to get each task done. Generally, one does not just take a meaningful exam or turn in a term report. One goes through a series of steps, smaller tasks, to get the main task done. For example, in taking an exam, one generally studies for it. Studying for it might mean reviewing class notes, doing research in the library, reading suggested publications or rereading the chapters covered in the class text. On the other hand, if one needs to write a term paper one may need to develop a topic, create an outline, maybe do some research – maybe on a number of aspects of the topic – develop a draft, and edit and finalize the draft. Many of the things we do, do not need a <u>formal</u> plan. We know them well and can take care of them without any formality. However, for the less well-known or more complicated tasks we should be sure that we have the approach well in mind. Further, we should have in mind about how long we need to spend on each of the steps to get the job done and when we will do them. We still might not have to write anything down, but we should be sure that what we have in mind will get the job done and get it done right. For some tasks a <u>written</u> plan is a big help - if not a necessity. Steps may break down into sub steps and so on, each with an estimate of the amount of time that it will take to get that step or sub step done. Further, the order in which the steps or sub steps can be accomplished is important. Don't get caught in the trap of not being able to accomplish a step because of some unforeseen situation – the book you need is nowhere to be found or you just can't seem to solve a key math problem and it is too late to call your friend. Always have alternatives – what other steps or sub steps can you do or maybe feel like doing when the one you planned to do next doesn't seem to be working for you. Further, you need to maintain flexibility as your plans may change over time. For example, that term paper you have to write for your physics class. Well today the instructor said to keep it short, no more than 3 pages, as he doesn't like long ones. So now some of the time you were going to spend on the term paper, since you had a much more elaborate one in mind, can be devoted to other activities.

OK, so now you have events and some approaches to completing at least some of the events. This is true not only because some of the requirements or events themselves may change, but because the steps one has in mind to complete the requirements for the event will often get refined over a period of time. So you may not have all of the details when you start executing your plans. That's life, we put down what we can think of today and seek to increase the level of detail as time moves on. The next step is to schedule the steps into a plan. As noted above, it is good to have in mind at least a basic schedule. Things that you know you can do in specific days in the future. For example, you may be able to accomplish a number of steps on different tasks with one visit to the library that you have planned for this Friday. However, you need to get used to having a daily plan. The first step is to determine what needs to be done today. Another step is to determine if time were available – what would you do with it, what can you get done today rather than leaving it for tomorrow? Maybe the test you planned to study for today is far enough away that no work is needed on it today. But maybe some research is required on the term paper, maybe extra credit, or homework that is not due until the end of the week

It is important to remember that of all of the things that one can buy or obtain, time is not one of them. Further, as the semester moves along, time becomes more and more precious. The message here is to plan each day. When I entered college I began to look at days as having three time periods – morning, afternoon and evening – with about 4 hours each. For example; 8 to 12, 1 to 5 and 6 to 10. Further, I found that I did better at creative things, like writing, early in the day so I tried to schedule them in the morning session. Things that I looked forward to doing, just wanted to do, were easy to do or were just dog work I generally tried to schedule in the evening because that seemed to be my least productive time period. If I allowed myself to say read a book that I wanted to read earlier in the day I might never get to the tasks that really needed to be done. Other tasks, like solving hard engineering problems I generally tried to do in the afternoon – a tradeoff between not using my creative time and not waiting until the mind was exhausted. Thus, my days tended to start with creativity, move to problem solving and wind up with the routine – and they still do. Your work cycles may differ, but the three time period approach is still a good framework to work with.

Like I said above, time is one thing that we can't buy more of. Smart people use it wisely. To be efficient you need a plan – it doesn't matter whether you are a student, a worker or a manager. Doesn't matter whether you are trying to get things done at work or at home. I even plan my vacations so that I know what I need to take along or what arrangements I need to make to do what I want to do while I'm there. There is one difference here though – if I'm not following my plan, but I'm having a good time I just don't worry about it. When on vacation the object is to have a good time, not to follow the plan. My vacation plans are not formal. Most of the time I don't even write them down. But it does help to say "I think I would like to do this and that and maybe I could do this on Tuesday and that on Wednesday…". Even a simple plan like that gives me an idea as to how long I might want to be at a certain vacation spot as well as the reasons that I chose to go there in the first place.

So, when you get up in the morning, and before you actually start to work, spend a little time determining what you must get done that day and what else you would like to get done during the day. For example, perhaps today you really need to get started on the research for a term paper and study for an exam you have to take tomorrow. Further, you would also like to get the outline for an upcoming speech started, but it doesn't have to be done today. Then, determine how long you need to spend on each task and make out a schedule for the day's activities keeping in mind what must get done and what would be nice to get done. Further, also keep in mind that it's important to move all of the important items along and not to let any of them get too far behind. Further, don't get so wrapped up in one item that you neglect the others.

Let me introduce a concept that I have used effectively since my early days in college. I call it Time Slices. First, as I noted above, I divide the day into three main periods; morning, afternoon and evening. These are the major Time Slices. I reserve the night for sleeping – well at least most of the time. Each period contains about four hours. Divide these four hours into whatever Time Slices that you need to get each item done. Further, as already discussed, determine what part of the day you are at your best doing different types of tasks – for example, you may be like me - more creative in the morning. Maybe that's your time for writing or developing ideas. Afternoon perhaps a mix. Evening maybe dog work as the mind may be tired. I like breaks in between the periods to eat, do other things and to give my mind a rest, so to speak. Actually, only the conscious mind rests. The subconscious works all of the time[7]. Sometimes when the conscious mind is at rest or not heavily tasked, the subconscious mind will provide answers to questions that the conscious mind has posed some time before – like in the middle of the night or when one is out for a walk. So sometimes it pays to relax and let the subconscious mind do its thing. Later I will provide more on the subconscious.

Time Slices can be whatever is needed to get a task done. A slice can be 15 minutes, 2 hours, 4 hours, it doesn't matter. However, when possible I like to have some flexibility in my plan. This is because from time to time I will run into a situation where I'm really doing well on a topic – often with ideas that I don't want to forget – and my schedule says that I'm out of time for that topic. Usually a quick look at the schedule allows me to move things around so that I can at least finish the particular aspect of what I am doing and not lose the thoughts that I have on the subject.

When things are new or changing it's a good idea to actually write down the plan for the day. Then carry it

---

[7] ibid, "The Human Computer"

with you and mark off the items as you complete them. Sometimes the schedule may become so repetitive, at least for short periods of time, that you have it so well in mind that there is no need to write it down. However, always be sure that you do have a plan and that you do follow it. It will be good training for the rest of your life as well as essential in being efficient in college. Figure 1 shows an example of a plan for a day.

g. Never Give Up

There are times when it is appropriate to give up but they don't generally apply here. For example, I once had $400 worth of damage done by a renter to a house I owned. I struggled to recover the loss. When I realized that it was going to cost $2,000 to recover the loss I gave up. However, what I'm talking about here are things like turning an incomplete exam in early because you have not been able to think of the answers or saying that you can't do something without at least trying it.

h. Reward Yourself

Bosses give out rewards for hard work. Now you are the boss, so be sure and reward yourself. It has been said, by some military commanders, that a warfighter will give his life for a small piece of cloth – a ribbon. Well, let's not go that far, but stop and think for a moment as to what motivates you or might motivate you in a more benign environment. For example, studying day after day can get somewhat old, especially when you expect to do it for at least four years. Here is where rewards can play a part. For example, you may really

Morning ....
8:00	Breakfast & plan day
8:30	"
9:00	Term paper research in library
9:30	"
10:00	"
10:30	"
11:00	Develop annotated term paper outline
11:30	"

Afternoon ....
12:00	Lunch & short walk
12:30	"
1:00	Math class
1:30	"
2:00	Math homework
2:30	"
3:00	"
3:30	"
4:00	Physics class
4:30	"
5:00	Dinner & read newspaper & mail
5:30	"

Evening ....
6:00	Physics homework
6:30	"
7:00	"
7:30	"
8:00	Math homework
8:30	"
9:00	"
9:30	Rewrite notes from classes taken today
10:00	"
10:30	Short walk
11:00	Read favorite book
11:30	Retire

Figure 1 Example of a Plan for the Day

yearn for a walk in the cool of the night, and even though it would not take all that long, time is short and you still have homework to do. So, you promise yourself that if you can adequately answer all of the homework questions by nine o'clock, you will reward yourself with a ten-minute walk.

Now you may confuse your subconscious if you don't do the work, but go for a walk anyways. However, this is not a black and white situation. Maybe you really tried, but it was not possible to accomplish the task in the time allotted. Maybe there is a trick to one of the questions. Maybe the walk would actually be a good thing to clear the mind – so to speak. The way to look at this is you are the boss. Did you fail because you goofed off or did you really work, but the task was too much to be accomplished in the time allotted? OK, so it's hard to do, but it's not going to be any easier to do with your children or your employees, so you might as well get started working on it right now and who better to practice on than yourself. Here you have the advantage. You know the truth – at least as best as it can be known. You will rarely have that in other instances. My father had an interesting saying and over the years I have heard it from others. "Believe nothing you hear and only half of what you see." There is a lot of truth to that. There is also a lot of truth to the concept that you can get a horse to move forward if you hold a carrot in front of its face. However, if it never gets the carrot it will soon tire of the effort. So when you demand a lot of yourself, be sure to reward yourself for a job well done.

3. Talk To Your Professors – With Purpose, Understanding And Tact

This technique might seem obvious to some and ridiculous to others. Many young students that I have talked with seemed to want to blame their professors or others for the bad grades that they got or were about to get. Some wanted to blame "other" factors. While the professors do make up the grades, the students do generally *earn* them. On the other hand, I'll include a discussion of some counter examples (read unfair) later in this section.

It is always good to get to know your professors - in a positive way. This is very important especially if you think there is any chance that you might not be headed in a direction that will result in the grade that you are working to get. I'm sure that you have heard of the expression "It ain't over until the fat lady sings" which is often attributed to Yogi Berra of baseball fame. Well don't wait until you hear the music to seek help. On the other hand, don't give up until you do.

You might think of it this way. If the professor does not know you or what you are trying to achieve, you are on your own. However, if he or she does know what you are trying to achieve you might find a whole different situation. I want to specifically emphasize that I'm not talking about buttering up the professor or any of the other things that may come to one's mind at this point. I'm strictly talking about guidance.

I got to know most of my professors. Some encounters were more rewarding than others. Some professors gave me much more work to do than I bargained for. Regardless of what they demanded as far as additional work was concerned (and none ever suggested anything else), once started I never backed off. If you are adverse to hard work, you might want to approach this technique with a bit of caution. However, if the thought of a bit of hard work doesn't deter you, I will say that in nearly every case I did raise my grade, nearly always to an A. Let me give you some examples. But before I do, let me make one point clear. If you ask a professor for his or her advice and then don't take it, you are probably worse off than if you had never asked. On the other hand, my experience indicates that most professors believe that they are effective teachers. Thus, when a student comes along that appears to be able to learn the subject and is really making a concerted effort to do so, but is just not getting it, it is bound to generate a dilemma in the professor's mind – whose fault is it - teacher or student? I'll leave the study of this as an exercise for the reader.

This technique is not all that hard to apply, but it seems very hard to generalize. Each situation is somewhat unique and requires some thought on what one is trying to accomplish by talking to the professor, when to talk to the professor and how to make sure that one accomplishes the task that one is setting out to do. I know of no formula to bring all of this together, so I'll provide some examples to illustrate how I employed the technique.

a. Two Minutes To An A

At the start of what was called an *introductory* physics course, the professor explained that there were going to be four exams during the semester and a final. He said he would disregard the lowest of the four exams and, for students that wished to try to improve their grades, he would give a make-up exam that could be used to replace any of the three exams that were to count. Even though I had my Bachelor's and Master's degrees by this time, this was a required course for me. I didn't mind. It was a chance for a good review and an easy A. However, I got C's on the first two exams even though I studied the subject well. After exhausting all of my ideas on what I was doing wrong, I decided to consult with the professor.

The professor was an extremely nice middle-aged man, very happy to help a trying student. I quickly reviewed why I was there and he asked to see my exam papers. We looked at the first problem on the first page. I showed him how I worked it. Then he asked why I didn't use the formula provided at the bottom of the exam. I was in shock. I couldn't believe it. I had never looked that far down on the page. I had never looked beyond the last question. There was no need - until the professor showed me the formulas and other information that was provided there. He placed the material there so we didn't have to memorize the stuff. Not knowing this I had committed everything to memory and as where my method was valid it didn't result in the specific answer being looked for. I knew right then that I was going to get an A in the course and I did. I got 100's on the next two exams, on the extra exam the professor offered and on the final. An added benefit was that I didn't have to spend all of that extra time memorizing the formulas and other data needed for the exams. And the change took just about two minutes with the professor.

b. In The Nick Of Time

For reasons beyond my control, I arrived at MIT at the start of the Spring semester. My doctorial committee chairman saw all of the math I had already taken and suggested that I might be interested in taking a course on Artificial Intelligence (AI), and I was. It was a super interesting course and the professor really knew the subject. The fact that there were no homework assignments and no exams were forth coming didn't bother me at first. However, after taking the midterms for all of the other courses, I began to get concerned when one did not appear in the AI course. Further, I realized that the AI professor had not even mentioned how grades were going to be earned in his course. So I thought I best schedule a meeting with him.

In the meeting he said, "As I told you last semester..." "But," I said, "I wasn't here last semester". It turned out that I was in the second semester of a two-semester course. In the first semester the class learned a special computer programming language. This programming language was designed to facilitate the writing of heuristic programs or programs that mimic the way humans do certain tasks. The second semester classroom lectures covered typical heuristic programs. The homework was to write such a program using the special language learned in the first semester. The grade for the second semester was to be totally based on the program the student wrote.

The professor realized right away that I could not learn the programming language and write the required program in the remaining five weeks of the semester. He suggested that I consider writing a paper on the subject and that my grade be based on it. It sounded like a reasonable option - perhaps my only option, but what could I write about that would get me an A in the course. An A that I really wanted to get since this was the first semester of my Doctorial program and I wanted to get it started on the right foot.

I decided to do a rather simple heuristic experiment. It didn't seem nearly so simple back then, when I was just learning about the world of heuristics, as it appears today. The experiment took about three to four weeks to complete as it involved about 25 people, none of whom were at the university. In fact most were in the order of 1000 miles away and I had to correspond with them by mail. There was no such thing as email back then.

I wrote a paper describing the experiment and the results I had obtained and submitted it to the professor before the end of the semester. Much to my surprise and to my sheer happiness, when the grades came out I had received an A. I saw the professor later and discussed the paper and grade with him. He said that he gave me an A because I was the only one in the whole class, which was so large it was taught in an auditorium, that had done an experiment to show a heuristic effect. All of the other students, each of whom had written their version of the requisite computer program, had assumed that the effect their program demonstrated was heuristic.

The lesson here is find out early what the professor expects. The earlier you learn the rules of the game the more time you have to do what is necessary to get the grade for which you are working. However, from time to time things happen, as they did in this course. Then, one needs to seek appropriate help.

c. A Semester Of Hard Work

College is unfair. I have often heard this complaint from students. As I noted above, life at times seems unfair. Why would one expect college to be any different? It's our job as parents to prepare our children to be able to deal with this unfairness. I didn't say correct it or change it, but to deal with it. Sometimes, however, in correcting unfairness to one's self one inadvertently causes unfairness to others. This is one of those cases and I never saw it coming. Even if I had seen it coming I don't think I could have done anything about it.

This was a two-semester course in electrical engineering. In the first semester there were four one-hour exams during the semester and a three-hour final. I got 100's on each of the hour exams and a 97 on the final. I made a slide rule (calculators had not made their appearance yet) error on one problem and a simple addition error on another. One's final course grade was totally based on the results of these five exams. I got a B.

When I saw the B ... well I just knew that it had to be a mistake given my grades on the exams. I went to see the professor. His simple explanation was that I was not perfect. Out of the 130 or so students in the class only one received an A. As you can imagine, for a student who worked to the extreme to get all A's, receiving a B with an average grade above 97 for the semester was very hard to take. It troubled me a great deal, but some things we just have to live with. Further, that was the good news, things were about to get worse.

The second semester was taught by the same professor. It was a required course and he was the only one teaching it. I had to take it. The class opened with a bang; perhaps a dud might be a better description. That's because I got a D on the first hour exam. Worse yet, I got an F on the second hour exam. Way past time to have another meeting with the professor, as I'm sure you would agree. As you can imagine I was not looking forward to the meeting since I still had the B around my neck from the first meeting. He had a consulting job in addition to his teaching job, so he only came to school when he had to teach a class - unless someone had arranged a meeting with him, which I did. His solution to my current problem was to tell me to work out every problem in the back of every chapter of the book we were using that semester. We're talking engineering problems, each of which can easily take hours to work out. He said he would be in his office every Thursday afternoon to go over the problems that I had done that week. This was very interesting to me since I knew about his other job. Spending that kind of time with me must of cost him big bucks, but I met with him every Thursday afternoon of every week for the rest of the semester. I really worked, but I didn't see an alternative. If I didn't do it I wouldn't even get a B for that semester. On the other hand, by working through each of the problems with him on the Thursday afternoons for the rest of the semester, I not only understood the subject better, I ended up with his approved answer for each of the problems.

By the end of the semester I had worked every problem in the whole book. The notebook I kept the answers in was huge. Then came the final. Final exams at the university covered two weeks. The university's schedule allocated three hours for a final exam, but there were no constraints placed on the professors. The professors simply told us how long he was going to give us. I took many eight-hour finals. It was not unusual for a final to be from 8 am until midnight. It was not even unusual to have to leave one final to go take another and then return to the first one.

All of this professor's exams were open book. You could bring anything except another person. His exams were such that the book, notes, etc. didn't usually help much if at all. On this final I don't remember him even setting a time limit. What was distinctive about this exam was that it consisted solely of about 20 problems from the back of the chapters we had covered during the semester. We're talking days of work not hours. And there I was with my notebook with not just the answers to all of the problems; I had the professor's approved answers. It was an extremely easy test for me. I was done in a couple of hours. It took me that long just to copy the answers from my notes, which was well within the rules since it was an open book test. For everyone else it was an impossible test. No one else even came close to finishing. Of course I got the only A in the course that semester. Now it's not as bad as it might seem. Many of the other students knew I was working out the problems during the semester and going over them with the professor. I made no secret about it. It was so much effort no one else ever even suggested that they join me. One might think that getting that A should have made up for the B, but it didn't.

d. Hostile Professor

This was one of the most difficult courses I ever took. Not because of the course material, but because the way the course was taught, the way the students had learned to deal with it and the apparent hostility of the professor toward me from the opening moments of the first day of class. Further, it is also one of the best illustrations of the impact of talking to the professor.

This class was different from the moment I considered taking it. For one thing, it was the only course I ever took in the College of Business and I want to state very clearly that the way this particular course was taught and what happened in it has nothing to do with the way other courses in the college were taught. In fact, what happened here could happen in any class in any college in any university. Further, I have a great respect for those in business. Without them there would not be as much of a need for engineers.

There is a premise in engineering that if you are a good engineer you will end up in management. I took this to heart in selecting my courses and decided to take this course in Personnel Management. The problem was that registration for students in the College of Business occurred one day before registration for students in the College of Engineering. By the time I got to register the Personnel Management course was always full. So one semester I decided to get a bit creative (we'll leave the how out to protect the innocent) and I secured a spot in the class. I spent the rest of the semester regretting what I originally thought was a clever move.

The first class started off with the professor calling the roll, which I found a bit strange since professors in engineering didn't do that. The professor never looked up from his roll book until he got to my name. He stopped, asked me to identify myself and then asked me how I got into his class. It seemed that I was the first engineer he ever had in his class. Since I had gotten into the class in a nonstandard way I wasn't sure what to say. At that point I still really wanted to take the course and I didn't want to get those that had helped me register for it in any trouble. Before I could answer the question, however, he began to tell me, and the rest of the class, that he had gone to the College of Engineering and looked up my record. Further, he stated that no one deserved the honor point average[8] that I had and that he was going to prove it to me. At this point I clearly began to question the cleverness that got me into the class.

After the reading of the roll, the class really began to go down hill, at least my part of it. The professor constantly made me do extra things - one might say he picked on me. It started with the professor saying, "Class open your books to page 26 (or some other page) and Lane start reading". At that point in my life reading out loud was something I really didn't like to do. I got so much practice that semester I think he cured me. Other times it would be that he needed two volunteers to do some project. He would say, "Lane and who else will volunteer?" Although all of this was very embarrassing

---

[8] The average of all of the final grades one has received in all of the courses that one has taken up to that point in time.

to me, the professor was so overbearing I never objected. I really thought I needed this course.

The real bell ringer came with the first exam. It was an hour long and a good part of the class turned in their papers after about 20 minutes. I felt sorry for them. Silently to myself I said, "You guys should have studied!" I stayed the full hour. In engineering I had learned not to expect to finish an exam, especially one with essay type answers. I reviewed my answers before turning in my paper. I had answered all of the questions and I was pleased with the answers. At the next class the professor read the grades; something else I had not experienced in engineering. The grades ranged from about 97 to 100 except for Lane. He got a 57. Time for a chat with the professor.

My answers were nearly direct quotes from the book, but not quite. Like where the book has the word "to" I had the word "for" or something as minor as that. I claimed the meaning was exactly the same. He insisted that I was inexact and the 57 stood. At that point I began to wonder how others in the class could, in as little as 20 minutes, write down the answers to all of the questions exactly as stated in the book on a closed book exam.

The results of my poking into the problem were horrifying. The situation I found myself in was the following. There were three classes of Personnel Management taught by this professor. A Monday/Wednesday/Friday morning class, a Monday/Wednesday/Friday afternoon class and a Tuesday/Thursday class. I was in the Tuesday/Thursday class. Exams were given on Monday and Tuesday or Wednesday and Thursday. Each class was given exactly the same exam. Questions on the exams were from exams given by the professor in the past. Files of these exams existed in organizations around the campus - none of which I had the time, money or inclination to join. As noted above, I generally carried 19 to 21 hours a semester, was married and by this time had a child to help take care of. There was no time for social organizations. However, it didn't take long to figure out that at least some, perhaps most, of the students in the Tuesday/Thursday class were coming in with their exam papers already filled out.

There were three other one-hour exams in that class. No matter how hard I studied the results were about the same. I don't remember another student ever getting a grade below 85. I never got above 57. Time for an extreme strategy, wouldn't you say?

However, I had not met this type of extreme condition before and as the class ended that semester I was not quite sure how to proceed. However, I did have one rule. Don't ever give up. So I decided to have one more chat with the professor. I decided that it would be a positive one - I never planned nor had a negative chat with any of my professors. No matter what they said or what they gave or told me to do, I made sure the meeting was positive, friendly and upbeat. I decided that I would continue along the lines of previous discussions, i.e., I really was working very hard to improve my grades. However, this time I would add, in a very subtle and hardly noticeable way, that I would bet that if he gave a final exam made up of questions that had not appeared on any of his previous exams, I would be one of the few people that would be able to pass it. I decided that I would weave this in the conversation without ever saying or even implying that he used old exams or at least questions from old exams.

We had a good meeting. When I brought up the subject of a final with all new questions he never said a thing. He didn't even blink. Actually, what I said took so few words and fit so well into the conversation that I wondered if he had even heard me or understood what I had said. I thought about repeating it, but decided against it. By then I knew his background. He was a smart man. He had not argued the point about the 'old' questions. I bet that he was not responding because he was thinking about the implications of what I had said. Forcing him to respond could have detrimental results.

A week or so later I took the final. A three-hour endeavor. It didn't seem much different to me than any of his other exams, just longer. The grades were posted on the professor's office door in the College of Business a week or so later. As I remember there were about 135 students in the three classes so it took a bit to find my name. The grade was an A. What a happy surprise. Then I began to examine the rest of the grades. I was shocked at the large number of C's and D's. And there were many F's as well. I counted the A's. Only 9 including mine. It was clear what the professor had done. He tested my theory. That is, he gave an exam with new questions and it sunk the class. I had a feeling that when the rest of the class saw the grades there were going to be a lot of very unhappy, if not very angry, students. However, I did not hang around to confirm this.

e. Betting On The Final

A number of times I found myself finally beginning to understand the course as the semester drew to a close. This usually gave me the feeling that if I reviewed all of the material in preparation for the final exam, I would do better on the final than I had done during the semester. These were usually classes in which I had a B or C average prior to the final exam. In these cases I always went to the professor and asked if he (it never occurred with a female professor probably because I didn't have that many) would base my whole grade for the semester on the final exam. Not one professor ever refused. Of course, that arrangement put a whole new level of stress on preparing for the final exam. Interestingly, I never failed to raise my grade. In only one class can I remember only getting a B. In all of the other classes I raised my grade all the way to an A.

* * * * *

However, meeting with the professor didn't always work, at least not in the way I had hoped, nor did it always seem like the thing to do. Let me give you two examples; *A Flat No* and *Beyond Hope*.

f. A Flat No

This situation occurred in an engineering drawing class. I related well to the class and to the professor. It was mostly a hands-on class. It was a subject that I really liked and I really enjoyed the class. We spent most of our time in class drawing. We received a grade on each drawing. By the end of the semester we had completed a large number of drawings. My course grade at the end of the semester, prior to taking the final, was nearly a perfect 100. All of the finals for my other classes occurred early during the first week of the two-week period in which the university had scheduled the final exams. The engineering drawing class final was scheduled for near the end of the second week. My finances needed a boost so I had lined up a summer job, which with 35,000 students on campus was not easy to come by. The only problem was that I needed to start right away.

I decided to meet with the professor to see if, based on my grades up to that time, he would let me skip the final. Although it had never happened to me I had seen it granted to others in other classes. The professor said, "NO!" and the tone in his voice and the look that he gave me when he said it told me that he was not going to discuss the matter any further. I was shocked, partly because, in terms of grades, I was his best student. I had always had a good rapport with him and I wasn't just trying to skip the final, I had a job waiting and I really needed the money. He knew I had a wife and a small child.

I thought a minute and then asked what he would do if I didn't show up for the final. Ooh, bad news. He said he would give me a grade of 0 for the final exam. I thought another minute. "What kind of grade would that give me for the semester?" I asked, almost not wanting to hear the answer. He dug out his record book and made some calculations. "An A," he said. "Are you sure?" I asked. He said he was and I said, "See you around". I didn't take the final. I went to work instead. I sweated for the weeks it took for him to post the grades. After all, I had no proof that he said he would give me an A. Not even a witness. But in the end he did give me an A.

g. Beyond Hope

I hesitate to discuss this next one since I don't want to discourage anyone from utilizing this technique. On the other hand, I don't want to raise expectations unjustly high so I'm including it for balance just so you know that I didn't win them all. As you read the situation that I managed to get myself into you will probably come to the same conclusion I did – I was lucky to get a C, even if it was the only one I received in college.

This situation occurred in an advanced physics course. By this time I had nearly four children and I was pushing hard to complete my course work on my doctorate. I was required to take the course, but as it turned out, nobody said I had to take its prerequisites. So I took the course. Everything was wrong. It was taught in a building across the campus from the course I was taking in the preceding period, so I couldn't get there early enough to get the seat I discussed above. Worse, it was taught in an auditorium which by the time I got there was so full I had to sit in the last row. I could hardly hear the professor. It probably did not matter all that much as he spoke in broken English which I could barely understand even when I could hear him. And it got worse.

As the semester got under way I found out that there was no textbook for the class, no class notes and no references. Even worse, it turned out to be the second semester of a two-semester course. So in explaining solutions to problems, the professor would only go part way and then say, "... and the rest of it is the same as we did last semester". Of course I had no idea what he was talking about since I had not taken the first semester.

I did try to talk with him a number of times after class, but as you might imagine with an auditorium full of students trying to get to him and me starting from the back of the pack, i.e., the back row of the auditorium, I was not as successful as I would have liked to have been. Further, when I did talk with him I found him almost impossible to understand. It wasn't just his broken English. He had only recently come to the U.S. and to say that he had not yet become accustom to our ways would appear to be a gross under statement of the facts. Having traveled to his part of the world while I was in the Navy, I could appreciate his difficulty. After all he was speaking my language, although broken. When I had been in his country I was often reduced to communicating with hand signals.

Further, I realized that my claim to be a good student was on weak ground. I had skipped the prerequisites. It was not unthinkable for him to force me to take them. I decided a meeting with him was not in my best interest. I decided to work hard and accept my fate. It resulted in a C, the only C I ever received in my college career. But, given the situation, it could have been worse.

The situations above serve only as illustrations. It's what we can learn from them that is important. First, I wouldn't want anyone reading the above to think this approach works without fail. Professors are busy and if you are going to schedule a meeting with one of them, you should take that into consideration. You want to appear as a student who is trying to improve and not a pest or a waste of time. On the other hand, if you do bond with the professor, you do raise the stakes. Consider the following. Most professors are there to teach. Most believe they can. Then, once a professor identifies a student as capable of doing better, sincerely interested in doing better, and worthy of the professor's time to help him or her do better, if that student follows the professor's guidance, but fails to do better who has really failed? Another way to look at this is, can the professor, in his or her mind, really let such a student fail? This is in no way a reflection on the profession of teaching; it is a reflection on the fact that teachers are humans.

I wouldn't want anyone who reads the above to think that professors are pushovers. They are not. In general, they are smart and they have heard about every excuse on the planet. I can attest to the excuse part since as part of my education I was encouraged to teach at two universities. Granted my tenure as an instructor was short, but you get to hear a lot of excuses even in a short period of time.

On the other hand, most professors consider themselves to be teachers. Most are very sincere and very human. That is they are there to help students learn. The problem a professor faces is finding those students who really want to learn as opposed to those who are just trying to find an easy way to get through the course. So, if you decide to have a meeting with one of your professors, be sure that you are sincere and that you are willing to put extra effort into the class. Perhaps a great deal of extra effort.

4. Power For Life - Employing Your Subconscious

This is one of the most profound techniques one can employ. If you learn to master this technique you will find a capability that you can employ in many ways for the rest of your life. Few people realize the extent to which the mind controls the body[9,10]. I have never met a student that even considered training the mind, beyond the normal studying, to help improve his or her grades. Once I stumbled upon this technique and started using it I never mentioned it to anyone, other than my wife, until I was well out of school. No sense helping out the competition – right? Then, when I did start talking about it I met a range of reactions. Some people actually tried it. Others found the idea too hard to believe. Still others actually went to the trouble to check it out. One student asked one of her professors about it. A psychology professor I believe. The professor told her that she, the professor, agreed it would work. The student was really taken by the professor's response. I don't think the student believed me, but she certainly believed her professor.

What we are discussing here is the subconscious mind as opposed to the conscious mind. Most people have heard of it in some fashion, but few seem to know much about it and like I mentioned above, no student I ever talked to ever mentioned anything about trying to employ it to improve their grades. The first question might be, why even consider using it? Think of it this

---

[9] ibid, "The Human Computer"
[10] ibid, "The Human Control System"

way. If you got paid by the hour and you had a robot that you could train to help you and with the trained robot you could get twice the work done at no additional cost to you, why would you not want to employ the robot? You would probably jump at the chance. Now if, with very little effort, you could arrange it so that you had twice the brain capacity to apply to taking exams you would probably jump at that as well. Well that's essentially what we are talking about here. First however, a warning. I don't claim to be an expert on the brain. I do know that it has power far beyond what most of us use. I also know that it can have negative as well as positive effects. So, although I am willing to share with you what I know, I strongly encourage you not to experiment with the mind without appropriate help. I know of no negative effects of any of the training that I did with the exception of the one experiment I did as a young teenager. As a result I promised myself that I would never again give my mind negative training thoughts. Let me explain what happened.

I was in my early teenage years when one day I got to thinking about how the mind might control the body. I decided to try an experiment. It seemed simple, straightforward and innocent enough. I would just try to convince my body that I had broken my arm and that it really hurt. I picked a point in the middle of my right forearm and began to concentrate on how much it hurt. I did this for hours. I concentrated on the fact that it was broken and that it really hurt. Along about mid-afternoon, I noticed that my arm really did seem to hurt right at the point where I had been telling my mind that my arm was broken. With the experiment over I began to do other things, but my arm continued to hurt. I soon began to get worried that I might have screwed up something because I could not get my arm to quit hurting. It was well into the night before I could convince my mind that my arm was not really broken and that it really did not hurt. All of this left a deep impression on me. Don't fool with the mind, at least not in negative ways.

It was years before I returned to the subject in any significant way. It occurred soon after I started into college full time. I observed that sometimes when I took an exam, the answer or the approach to finding the answer to a particular problem popped into my (conscious) mind shortly after I left that problem to work on one of the other problems on the exam. I knew that these ideas were coming from my subconscious.

I got to thinking about this process. I knew that these helpful, and sometimes not so helpful, ideas that kept popping into my mind were coming from the subconscious. I knew it was a very powerful entity. I knew that it worked on problems, concepts, ideas, etc. in the background that is outside of the realm of the conscious mind. I also knew that it appeared to work in a very random way. For example, that street name I was trying to recall yesterday suddenly pops into my mind. I also had been told that when a normal, healthy person cannot remember something it isn't the fault of the storage part of the mind; that is, nothing is ever really forgotten. It's that the mind can't locate the information, at least not in the time period it has been given. Further, it seemed that the subconscious mind was able to locate things that the conscious mind couldn't. Sort of like short-term verses long-term memory. Everything might be in long-term memory, but relatively few things are in the short-term memory where quick access might be the guiding principle. This seemed realistic, at least to a budding systems engineer. That is, in very simple terms, if one was going to design a mind, there are two basic types of information that it would have to deal with; items that need immediate response and items whose response can be delayed. Thus, we need a *conscious* mind to control the real-time tasks associated with eating, avoiding objects while we are walking, driving, talking, etc. I include here the mechanical functions such as walking itself. We would not want to have this mind off doing extended tasks such as long memory searches for fear that we would walk into a post, crash our cars or just simply fall down every time we tried to recall something. However, some people seem to do some of this in spite of the current arrangement of the mind. Thus, we would want to have another mind (processor) working on these longer-term tasks in the background and, when an opportunity presented itself, to provide the result to the real-time (conscious) mind. This second mind is what we refer to as the subconscious mind.

I began to wonder if one could train the subconscious to work on exam problems in near real-time. The question was how to do it. How does one communicate with their subconscious? I decided to try the following approach. Actually, I believe it's at least one of the recommended ways to take an exam. But remember, I never really learned to study or take exams in high school. The approach I used was to read the question as I had been doing all along. If I knew how to work on it, fine, then I would do it. If not, then I would reread the question very carefully and deliberately to make sure, or at least try to make sure, that I fully understood the question to which I was seeking the answer. However, it is important to note that at the same time I was also providing a clear indication to the subconscious that I was giving it this task and that I needed an answer right away. I would then go to the next question and repeat the process. When I got to the end of the test I would return to the first question that I was unable to answer on the previous pass and repeat the process. I would continue looping until I finished the exam or time ran out.

I can still remember the first exam where I fully realized that the process was working. It was a number of months after I had started trying to develop the technique. It happened during a one-hour test in a beginning electrical engineering class. I followed the above procedure, but when I got to the bottom of the exam I realized that I had not been able to answer a single one of the ten questions. I remember the sweat beads popping out on my forehead as the thought of flunking my first college exam entered my mind. As I started down the questions on pass two I found I was able to answer some of the questions. The approaches and solutions popped right into my mind when I reread the questions. I don't remember how many passes through the questions it took but I eventually got to the point where there was only one question left. It was a "Derive the following formula" type question and I just could not remember how to get it started. I reread and reread the question. I pressed hard on the subconscious as time was running out. Finally, the subconscious responded with "Work it backwards on a scrap piece of paper and then transcribe it onto the exam paper". That was one of the most profound things I had encountered from the subconscious up to that point. It couldn't come up with the starting point either; at least not in the time I was giving it, so it figured out another way to solve the problem. At that point I began to realize how complex and powerful the subconscious really is.

As a result of this I realized that I was on the way to developing a new way, at least for me, to take exams. And yes, the other good news was that I got a 100 on the exam - on an exam that at first reading I had the feeling that I was going to fail. From then on I always used this approach in taking exams.

I observed several other things that day. First, it appeared that the subconscious could handle multiple tasks at the same time – at least I could give it multiple tasks and it would give responses back in a different order than the requests that I made. It was not clear if it was processing the requests in parallel or one at a time in an order that it had selected. I also learned, from the "work it backwards" response, that the subconscious is much more than a search engine – you might say it has a mind of its own. I also learned that I could command it to do things. Later I will share with you some other things that I have learned to *tell* the subconscious to do.

APPLYING THE TECHNIQUES

Now that you have read the techniques, how does one really apply them? Here I must remind you that although the techniques worked for me, I cannot guarantee that they will work for you. Further, if you choose to try them or any variation of them, you must assume responsibility for any results of your actions. On the other hand, if you do decide to try one or more of them, you can lay claim to any success that they may bring you.

Well, if you have gotten this far, I assume that you have agreed to assume all of the risks so let's continue. **First**, to think that one takes these techniques, plops down in a seat in a classroom and the infusion of knowledge begins to take place is a bit naive. Actually data does begin to enter the brain, but the understanding of the information contained in that data and the development of a capability to use it is not a passive activity – at least not for most of us. It takes real work – in some cases real hard work. And so does the application of the techniques. But the results can be astounding.

Second, attempting to apply all of the techniques at once is a great, but probably an impractical objective. To get the real benefit from most of the techniques one must master them – generally by working on them over and over. It is possible that one will find a variant of a technique works best for him or her. There is nothing to say that the techniques as described are perfect or even complete. If you can refine them so that they work better for you, then do it. If you can expand on them, then do that as well.

Third, each class is different. The techniques that might be helpful will differ from class to class. When you go to a new class, size it up. That is, study the class. See if you can determine which techniques might be helpful in that class. Study the professor. What is he or she trying to accomplish and how is he or she trying to do it. You know the subject that is being taught, or is supposed to be taught, but each professor's approach will be somewhat different. Maybe a lot different. Develop a game plan. As you continue to attend class continue to study it and use your increased understanding of the class to refine your game plan.

Fourth, one can start with those techniques that are easy to apply like *The Only Seat To Sit In* and *Check And Double Check*. Others like *Look Sharp, Feel Sharp & Be Sharp* and *Plan Your Study Time* should not take much more effort. For these, one just needs to develop the right habits. But these habits will serve you well even after college – yes there is life after college.

Talk To Your Professors takes a little more effort. On the other hand, as discussed in the techniques, it can have a pivotal impact. As noted several times already and repeated here because of its importance – talking to your professors does not mean to try to butter them up or get on their good side. I think professors are smart enough to recognize that type of approach and to discount it outright – maybe even with a penalty. No, what "Talking…" means is communicating. Communicating in both directions. If you need help – you are working hard, but just not developing the understanding of the subject you feel you should be – then you should seek the advice of your professor. It's important to be sensitive and tactful. Sensitive in the sense that there are good times to have such a discussion with a professor and there are not so good times. He or she may be very receptive at the end of class and he or she may not. The professor may prefer to have such discussions in private and you may as well. Further, when you ask for help be sure to at least try to use it. If it doesn't work have another discussion with the professor. Be tactful. To ask a professor for advice without a sincere intension of trying to use it may actually be worse that never asking. Finally, if you make an appointment to talk to your professor, be on time. Be prepared to tell him or her, in a few words, what the problem is, what you have done to solve the problem and the guidance you are seeking from him or her.

Last, and probably most important in college and beyond, is the development of the ability to direct one's subconscious – in a positive way of course. Here I can only encourage one not to give up. I know no way short of continued practice to develop this capability. The exam approach discussed above seems to have merit, but there may be better ways. I continue to work on some aspects of this, as I will discuss in the next section, but to get started I know of no better way than the exam approach. It has two advantages – first, you'll have plenty of opportunity to practice and, second, you will undoubtedly know when it kicks in.

EPILOG

OK, So Who Really Cares About Your Grades?

Now that you have read the book, let's look at this again. You have probably come to realize that a whole lot of people care. For starters, you care - you must if you read the whole book. The universities care; it's part of the process associated with providing scholarships and fellowships – most of my college tuition was paid by organizations that cared and every one of them looked at my grades. Your family cares. The company you go to work for will care. And the list goes on.

So let's try to put all of this into perspective and see how grades, and even the attempt to improve one's grades, might impact one's life. My first recollection of grades was as a high school student. As a high school student I essentially had no concept of the real meaning of school let alone grades. I know that my parents had some interest in me getting better grades, but I never knew why. I never gave much thought to the "why". I now know that high school grades are important especially if one wishes to attend college – even more so if it is a well-known college. However, college grades, especially the ones a student earns in the classes associated with the major for his or her Bachelor's degree – the ones the university publishes as a student's honor point or official grade point average - seem to be the ones that impact one's life the most. The whole world looks at them – for example, recruiters, personnel managers and program managers from prospective employers. However, so do the schools themselves. The professors, deans, and college administrators as well as upper management in much of industry and the government are all looking for those promising young students that deserve a chance to further their education. Whether at the Master's or Doctorate level, these students are considered to be the ones that are potentially going to make a difference in the world, hopefully to make it better. Further, the universities not only look at grades to determine who is going to get into the different programs, they also look at grades to determine who they are going to support with scholarships, fellowships and research grants. Further, some colleges allow top students a voice in what courses are necessary to complete their Bachelor's degree. This effectively increases the student's electives.

So the reasons for good grades are many – each of us has our own opinion about them. There is no universal opinion except perhaps that good grades are better than bad grades. I think it's fair to sum it up in the following way. I think that most professors try to give the students the grades that, in the professor's mind, the students earned. Getting a good grade makes one feel much better than getting a bad grade. Good grades make one feel good about one's self. Good grades beget good grades and may lead one to loftier heights. The opposite is often true for bad grades. Finally, your college grades are going to be with you awhile. In some cases they will be your calling card. They will be what makes the first impression about you – and you will undoubtedly want to put your best foot forward. On the other hand, let us not forget that although outstanding grades may open some doors it is what one does after the doors are open that really counts.

Added Capability

There is still another benefit that one might derive from working on and applying these techniques, and it may provide the biggest benefit of all – that of developing one's problem solving abilities. I noticed from the first time I went to work that I looked at problems a bit differently than many of my peers. Back then we thought they were neat ideas. Today we call the process "thinking out of the box". To me it seems very natural and I believe that it got developed, or at least refined, during the period of trying to improve my grades. Now there is both good news and bad news about this. The good news is that you might be able to solve problems that others cannot. The bad news is that you will need to grow a thicker skin. That is, people won't believe you, at least not at first.

Most of us have been told we can't do things at some point in our lives. Our parents probably started it when we were very small, but that's not what I'm talking about. If you have new ideas you will find people telling you that you can't do those things or that those things will never work. I have been told that I can't do things all of my life. Let me give you some real life examples.

The first significant instance that I can remember occurred as I was about to enter the university to start work on my Bachelor's degree in Electrical Engineering.

I met one of my high school classmates. He told me I couldn't make it in electrical engineering. Actually he was on pretty solid ground as no one in our high school class had made it in any engineering. His brother, who was two years ahead of us and very smart, didn't make it in mechanical engineering. His words certainly gave me cause to stop and think about what I was setting out to do. Since I had been in the Navy, which had taken me out of the country, I had not heard what had happened to my classmates that went off to college. The more I thought about it the more I realized that he had said "no one had made it". He hadn't said it couldn't be done. So I felt that I had to try and, well, I reported the result of that effort in the introduction.

You may have to get use to developing somewhat elaborate examples or demonstrations to prove your point. Let me give you a specific example. Within a few months of joining my first company I observed that a certain missile system in development went astray in about half of the test launches due to a programming error (each flight having a different error or set of errors) in its guidance computer. The company was responsible for writing the missile's guidance program. I was only indirectly involved, but I could see that the programming team was intelligent, dedicated and greatly troubled by these repeated failures. Although I didn't know it at the time, this problem had existed for a considerable period of time before I joined the company. I had an idea for a new type of computer programming tool that I felt would solve the problem. However, as I said, I was only recently out of school and on my first job as an engineer, so when I presented my idea to management I got told that the idea had already been thought of and that some of the best minds in the country had judged it impossible to solve. My supervisor believed in me, but his boss told us to 'get out and never come back'. After working at night at home for two months to develop a demo that showed that I could do it, I did go back. This time the boss believed me and gave me company funding to continue the effort as part of my normal work. It did stop the programming errors and the missile system became operational. Later the technology was applied to other computer programs. It became the basis for my Doctorial thesis, "Recomposition – A First Step In Computer Program Analyses". It was the start of the analyses of computer programs.

Temper Thy Eagerness

There is another aspect to this and I must admit, I still have not learned it very well. Sometimes you might think you can see how to solve a problem, but what you don't realize is that you haven't been given all of the facts. From time to time I have set myself up because of my eagerness to solve problems. When this happens one must look at the reasons others have failed and think about other ways to approach the problem – think "out of the box" – else you may not solve the problem either. An example of this happened when I agreed to do a task for NATO. Only after I had agreed to do it in 18 months did NATO tell me that there was a NATO committee that had been working on the problem for 9 years and was nowhere near solving the problem. As it turned out they were using the same approach that I had planned to use. Time for a new approach! Not only did I develop a new approach and complete the task in the planned 18 months, NATO used the results as the basis for one of their standard agreements – STANAG 5620.

The Good And The Bad

Reading this book you have probably gotten the idea that I believe in the power of positive thinking. Well you are right. There is a lot to be said for it. No, just because you think you can fly does not mean that you can fly, but positive thinking does go hand in hand with problem solving. To begin to think about solving a problem, especially one that no one else has solved, you have to start with the idea that there is a chance that you can solve it. Start with the idea that you probably can't solve it and you will very likely give up at the first point of difficulty. So get rid of those negative thoughts. Negative thinking will cause you to lose power. Positive thinking will allow you to gain it. When you think positively you move forward, you can feel it. Start by thinking, "How can I do it?", "What are the steps?", "Where do I begin?", "What is the first step?" Positive thinking will make you feel good, your body will act differently and your mind will respond.

Beyond Work – The Rest Of Your Life

Those examples are work related. However, problems occur in all parts of our lives. Let's take one from the more daily part of life – a part to which, I'm sure, many of you can relate. When I traveled all over Europe every month it didn't take me very long to realize that this was not a normal office job. For one thing, the bathroom was no longer just down the hall. While I tried to *go* every morning before I left the hotel, sometimes it just wasn't going to happen. I'm sure that everyone has experienced this. However, since I often moved from city-to-city or country-to-country every other day, I was often in unknown areas until I reached whatever hotel I was staying in that night. Sometimes desperation set in – I know some of you have experienced this. So I started telling my control center – my subconscious in this case – that if it didn't go right after breakfast it could not ask (read demand) to go until I got to the hotel that night. If it failed to go I would remind it of this as I left the hotel. Sometimes during the day the urge would begin, but I would tell it again. Over time it became very good at conforming to the rules. However, when I reached my hotel at night my room best be ready. To this day I still maintain that same control over it and I must say that it has saved me a lot of grief over the years.

Finally

First, let me restate a very important point about working to improve my grades. Although I had been told of its importance, just working to get better grades seemed a bit dull. Probably since I hadn't put much value on good grades in the past and the payoff for the grades I was now working for was down the road some place – a road I was yet to travel and a road that would take me into areas to which I could not yet relate. So I decided to make a game out of it. The object of the game was to get the best grade in the class that I could. Here I said *get* as opposed to *earn*. That's because, as I have noted in several places, what I thought I had *earned* and what the professor thought I had *earned* was not always the same. So to stay away from the question of whose definition of *earned* am I talking about, I changed the word to *get*. On the other hand, I don't recall ever getting a grade that was higher than what I thought I had earned. Also, I want to remind you of what I said in the beginning of the book - all of the efforts had to be fair, above board and available to all of the other students. Anything that would be considered unfair, inappropriate or unethical by the professor or the university was strictly taboo.

So, if I was taking six classes then I had six problems to solve and 15 to 17 weeks to solve them. And, I had a way to measure my problem solving skills – the grade which I received for the class. Thus, the game approach changed my focus from working for grades to solving problems. I could break the six problems down into many sub problems, which I could work on all semester. It made it much more fun and I believe aided the development of my ability to solve problems.

Second, I didn't say much about body language because I only began to recognize its significance late in my college days. The time I remember encountering it the most, I was taking a very important oral exam. I was at a blackboard in front of a group of professors seated around a table. One of the professors was asking me questions that were at the fringe of my knowledge. He had a reputation of being extremely tough and very intimidating, and he was living up to his reputation. I was answering one of his early questions and as I turned from the blackboard to the group I noticed that his head was moving sideways and he had a frown on his face. While both of these were very slight, they were clearly there. I quickly said something like, "...but it could also be..." and his face brightened and his head started to move up and down – again very slightly. He was actually and unknowingly telling me the correct approach. Of course I passed the exam even if the other professors didn't provide the same level of help. However, I never did get around to developing this into a technique. Perhaps others will more closely examine its utility and let us know the results.

Third, as you can see, these techniques have had a profound impact on my life. They also appear to have an impact on those with whom I have shared them. Will they have the same impact on your grades and your life? No one can say for sure. However, one can say that if you don't try to employ them they certainly won't have an impact. On the other hand, a halfhearted approach is probably equally ineffective. Thus, if you are going to try them, make it your best shot. Really work the problem. Make a game out of it. What do you have to lose verses what might you gain? A little bit of extra work to potentially gain a whole new way of life. Sounds like a winner.

Do you know:

- How grades impact your life?
- How grades come about?
- How to get better grades and have fun doing it?
- How to use your subconscious?
- The one thing every recruiter will remember about you?
- What college can really do for you?

This book answers these questions and more. Even better, it provides techniques to help you get better grades.

Visit MaxGrades at: www.maxgrades.com or contact MaxGrades at: support@maxgrades.com

To order additional copies - send $12.95 plus $2.00 tax and S&H to:

MaxGrades
P.O. Box 10494
Burke, VA 22009

Name _____
Address _____
_____
& Zip _____
_____

Please allow 4 to 6 weeks for delivery.